Research Trends and Challenges in Smart Grids

Edited by Alfredo Vaccaro, Ahmed Faheem Zobaa, Prabhakar Karthikeyan Shanmugam and Kannaiah Sathish Kumar

Published in London, United Kingdom

IntechOpen

Supporting open minds since 2005

Research Trends and Challenges in Smart Grids
http://dx.doi.org/10.5772/intechopen.77415
Edited by Alfredo Vaccaro, Ahmed Faheem Zobaa, Prabhakar Karthikeyan Shanmugam and Kannaiah Sathish Kumar

Contributors
Jafar Rasouli, Ahmad Motamedi, Mohamad Baseri, Mahshad Parsa, Dinh Hoa Nguyen, Tran Huynh Ngoc, Tatsuo Narikiyo, Michihiro Kawanishi, Maher Azzouz, Binod Vaidya, Hussein T. Mouftah, Fernando Vaca-Urbano, Manuel S. Alvarez-Alvarado, Angel A. Recalde, Félix Moncayo-Rea, Alfredo Vaccaro, Ahmed Faheem Zobaa, Antonio Pepiciello, Sandra Castano-Solis, Daniel Serrano-Jiménez, Jesús Fraile-Ardanuy, Javier Sanz-Feito, David Jiménez-Bermejo, Nithiyananthan Kannan

First published in London, United Kingdom, 2020 by IntechOpen
IntechOpen is the global imprint of INTECHOPEN LIMITED, registered in England and Wales, registration number: 11086078, 7th floor, 10 Lower Thames Street, London, EC3R 6AF, United Kingdom
Printed in Croatia

British Library Cataloguing-in-Publication Data
A catalogue record for this book is available from the British Library

Additional hard and PDF copies can be obtained from orders@intechopen.com

Research Trends and Challenges in Smart Grids
Edited by Alfredo Vaccaro, Ahmed Faheem Zobaa, Prabhakar Karthikeyan Shanmugam and Kannaiah Sathish Kumar
p. cm.
Print ISBN 978-1-78923-891-4
Online ISBN 978-1-78923-892-1
eBook (PDF) ISBN 978-1-83968-006-9

For EU product safety concerns:
IN TECH d.o.o., Prolaz Marije Krucifikse Kozulić 3, 51000 Rijeka, Croatia, info@intechopen.com or visit our website at intechopen.com.

Meet the editors

Alfredo Vaccaro received his MSc degree cum laude and commendation in Electronic Engineering from the University of Salerno and his PhD in Electrical and Computer Engineering from University of Waterloo, Ontario, Canada. From March 2002 to October 2014 he has been Assistant Professor of Electric Power Systems at the Department of Engineering, Faculty of Engineering of the University of Sannio. From February 2011 to December 2013 he was the Scientific Director of the Bureau of the Research Centre on Pure and Applied Mathematics at the Department of Engineering, University of Sannio. On October 2014 he obtained the National Scientific Qualification of Full Professor in Electrical Energy Engineering. Since November 2014 he has been Associate Professor of Electric Power Systems at the Department of Engineering of University of Sannio. He is the Editor in Chief of *Technology and Economics of Smart Grids and Sustainable Energy*, Springer Nature, Chair of the PES-IEEE Task Force on Enabling Paradigms for High-Performance Computing in Wide Area Monitoring Protective and Control Systems, and Chair of the IEEE PES Awards and Recognition Committee.

Ahmed Faheem Zobaa received his BSc (Hons), MSc, and PhD degrees in Electrical Power and Machines from Cairo University, Egypt, in 1992, 1997, and 2002, respectively. Also, he received his Postgraduate Certificate in Academic Practice from the University of Exeter, UK, in 2010, and his Doctorate of Science from Brunel University London, UK, in 2017. He was an instructor from 1992 to 1997, a teaching assistant from 1997 to 2002, and an assistant professor from 2002 to 2007 at Cairo University, Egypt. From 2007 to 2010, he was a senior lecturer in renewable energy at the University of Exeter, UK. Currently, he is a reader in electrical and power engineering, an MSc course director, and a full member of the Institute of Energy Futures at Brunel University London, UK. His main areas of expertise include power quality, (marine) renewable energy, smart grids, energy efficiency, and lighting applications. Dr. Zobaa is an executive editor for the *International Journal of Renewable Energy Technology*. He is also an executive editor-in-chief for *Technology and Economics of Smart Grids and Sustainable Energy*, and an editor-in-chief for the *International Journal of Electrical Engineering Education*. He is also an editorial board member, editor, associate editor, and editorial advisory board member for many international journals. He is a registered chartered engineer, chartered energy engineer, European engineer, and international professional engineer. He is also a registered member of the Engineering Council UK, Egypt Syndicate of Engineers, and the Egyptian Society of Engineers. He is a senior fellow of the Higher Education Academy of the UK and a fellow of the Institution of Engineering and Technology, the Energy Institute of the UK, the Chartered Institution of Building Services Engineers, the Institution of Mechanical Engineers, the Royal Society of Arts, the African Academy of Science, and the Chartered Institute of Educational Assessors. He is a senior member of the Institute of Electrical and Electronics Engineers. He is also a member of the International Solar Energy Society, the European Power Electronics and Drives Association, and the IEEE Standards Association.

Prabhakar Karthikeyan Shanmugam completed his BE (EEE) from the University of Madras, Tamil Nadu (1997), his ME (Electrical Power Engineering) from the M.S. University of Baroda, Vadodara, Gujarat (1999), and his PhD from VIT University, Tamil Nadu, India (2013) under the guidance of Prof. D. P. Kothari. He also completed his Postdoctoral Fellowship from the Central Power Research Institute, Bengaluru, Karnataka, India. He is presently with VIT University as an associate professor. He is a Senior Member-IEEE with 31 published articles in peer-reviewed journals and 77 works in national and international conferences. His area of interest includes deregulation and restructured power systems under smart grid environments, electric vehicles, and issues related to distribution system studies.

Dr. Kannaiah Sathish Kumar, PhD from VIT University, Vellore, India, has 14 years of total teaching experience and 9 years of research experience working in various research teams to develop new applications of evolutionary computing algorithms for solving various power system problems such as unit commitment, economic dispatch, emission reduction, smart grids, power system reconfiguration, and restoration. His current area of interest is studying interconnection problems in linkages of HVDC and HVAC (765 kV) transmission lines with existing 230 kV high-voltage lines, nano-additives for high-voltage XLPE cables, and optimization of smart grids. He is a member of IEEE, IEEE-PES, and SSI. He has published 51 research papers in different journals and conferences of international repute and authored a book on power quality. At present he is working for the School of Electrical Engineering as an associate professor. He is a reviewer for various SCI journals such as *Electrical Power and Energy Systems*, *ISA Transactions*, *AAI*, etc.

Contents

Preface

Today, power distribution systems need to face several critical issues, such as aging infrastructures, new resources to meet flexibility requirements, renewable power generators, congestion management, and reliability coordination. The major challenges facing these complex systems include balancing between resource adequacy, reliability, economics, and environmental and other public purpose objectives to optimize distribution resources to meet the growing demand.

In this context, the large-scale deployment of smart grid technologies could play a strategic role in supporting the evolution of conventional power distribution grids toward sustainable, flexible, and self-healing networks composed of distributed and cooperative energy resources.

However, to realize these benefits, several open problems need to be solved, because smart grid computing paradigms can drastically differ from the traditional architectures conventionally deployed in power distribution systems management. These differences derive mainly from the characteristics and penetration levels of the dispersed energy resources, the presence of controllable loads, the power quality constraints, and the difficulties in predicting and modeling user and renewable power generator dynamics.

From this perspective, a crucial issue is how to support the evolution of existing distribution networks from static hierarchical systems to self-organizing, highly scalable, and pervasive systems.

In this field, modern trends are oriented toward the employment of new control, protection, and monitoring techniques that move away from the traditional computing paradigms to systems distributed in the field, with an increasing pervasion of smart and cooperative devices. The large-scale deployment of these new technologies in power distribution systems could lead to more efficient task distribution among the distributed energy resources and, consequently, to a sensible improvement of overall grid flexibility.

This book is composed of eight chapters, which are focused on the most promising enabling technologies and methodologies for smart grids, addressing many relevant topics ranging from flexibility management to various control and communication aspects.

The large-scale deployment of these advanced techniques could improve the technical, economic, and environmental performance of modern power distribution systems by allowing a massive pervasion of dispersed generating units, increasing the hosting capacity of renewable power generators, reducing active power losses and atmospheric emissions, and improving system flexibility.

This book not only focuses on technological breakthroughs and roadmaps in implementing these technologies, but also presents the much needed sharing of best practices.

Alfredo Vaccaro
University of Sannio,
Benevento, Italy

Ahmed Faheem Zobaa
Brunel University London,
UK

Prabhakar Karthikeyan Shanmugam and Kannaiah Sathish Kumar
Vellore Institute of Technology,
India

Chapter 1

Introductory Chapter: Open Problems and Enabling Methodologies for Smart Grids

Alfredo Vaccaro, Antonio Pepiciello and Ahmed Faheem Zobaa

1. Introduction

Modern power systems are facing several challenges related to the transition from a traditional, fossil fuel-based, and vertically integrated architecture to a smart, sustainable, renewable generation-based, and deregulated system. Smart grid is the key concept that allows this transition and enables a series of innovative applications thanks to the integration of information and communication technologies into power systems.

Smart grids involve two-way electric and information flows across generation, transmission, distribution, and utilization systems, to improve their efficiency, sustainability, reliability, and resilience compared to traditional grids. The attribute "smart" reflects the layer of intelligence added to the power system that is able to *sense* power system's conditions, *interact* with producers and users, and *react* to any unexpected conditions.

Figure 1 describes the main differences between traditional and smart grids [1–3]. The concept of a smart grid was developed in order to reach a set of goals:

- ***Sustainability***: smart grids facilitate the introduction of sustainable and clean technologies, such as distributed renewable energy generators, into power systems.
- ***Monitoring***: smart grids guarantee real-time observability of power systems, thanks to the capillary distribution of smart meters and advanced sensors.
- ***Adaptability***: smart grids can adapt to different and evolving power system's configurations and promote the development of innovative applications.
- ***Resilience***: smart grids improve power system's robustness to disruption caused by natural disasters, extreme weather, and unexpected faults, enabling self-healing.
- ***Transparency***: smart grids guarantee secure and transparent communications and information systems, allowing customers to take keen choices and to proactively interact with the system.

In order to support the evolution of existing power systems from static and hierarchical networks to decentralized and self-healing systems composed by cooperative and self-organizing energy resources, a commonly accepted framework must be established. To this aim, in this chapter the most promising enabling technologies will be presented, and the possible research directions aimed at

IntechOpen

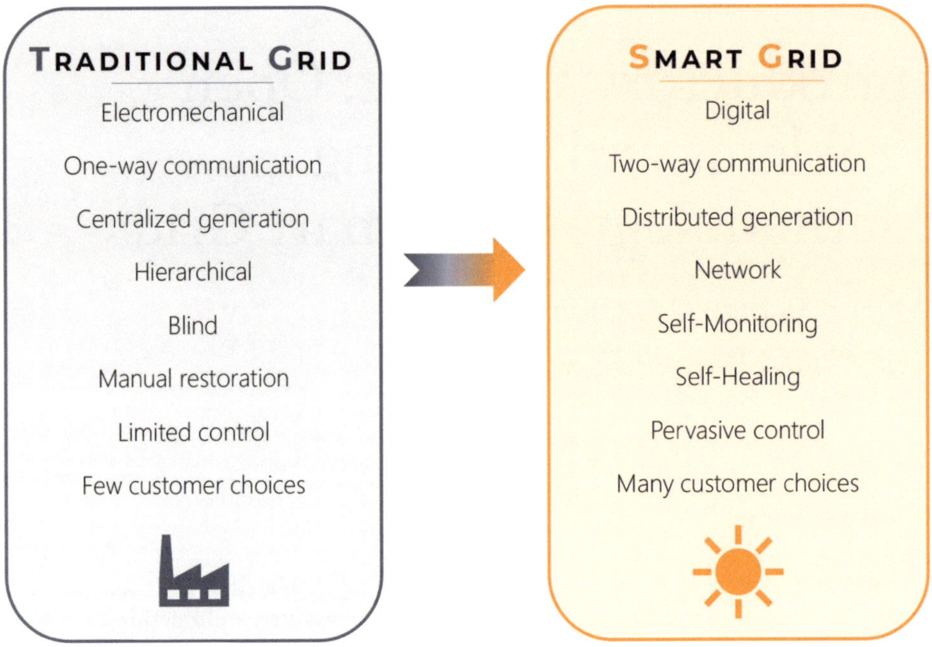

Figure 1.
Comparison between traditional grid and smart grid [1].

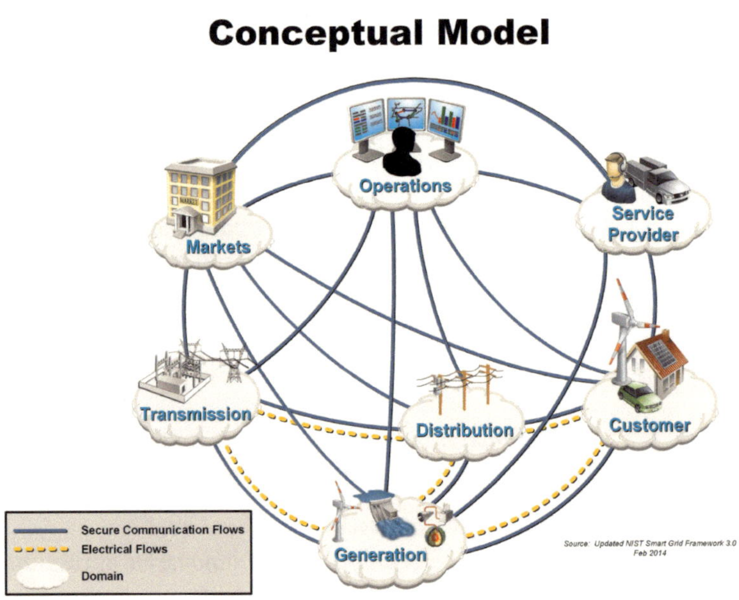

Figure 2.
NIST conceptual model [4].

addressing some challenging open problems, which could hinder their deployment in existing power grids, will be outlined (**Figure 2**).

2. Conceptual model

The most widely accepted smart grid's conceptual model has been developed by the National Institute of Standards and Technology (NIST) (**Figure 2**) [4].

It distinguishes seven different functional areas in smart grids, called *domains*. Each domain is characterized by a set of *actors* and *applications* that perform actions on energy and data inside the domain and allow the exchange of power and information between domains, by means of different *interfaces*, also called *domain gateways*.

The next sections describe the main features of each domain and the related key technologies.

2.1 Bulk generation

This domain collects all the actors and the applications related to the centralized generation of large amounts of power, optimally dispatched in order to satisfy the predicted demand. In traditional power systems, characterized by one-way power flows, this was the only domain in which power was produced.

The importance of this domain is not only due to the production of a high percentage of the total required power but also due to the ancillary services it offers in order to maintain stability and security of the whole power system.

This domain has communication interfaces to market, operation, and transmission domains, although it is electrically coupled only to transmission network.

2.2 Transmission domain

Transmission infrastructures allow the efficient transfer of electrical power from generators to distribution systems. Their main components are high-voltage power lines, substations, sensors, and protection systems. Their correct operation and maintenance is entrusted to transmission system operators (TSOs). The evolution of power systems toward smart grids is leading to the development of several innovative technologies to improve power transmission's reliability and efficiency [5]. The most impactful ones are advanced power electronic-based systems, such as flexible AC transmission systems (FACTS), that enhance controllability and increase power transfer capability of transmission systems.

2.3 Market domain

Market domain is where prices for power exchanges are established. Currently, electricity markets are changing in order to fully exploit the possibilities introduced by future smart grids.

Market domain is connected by communication paths to every other domain. It receives information regarding system's state and constraints by the operators and service providers and proceeds to dispatch generated power in order to satisfy the demand.

Communications between market domain and the other ones must be secure, reliable, and transparent and with low latency in order to correctly operate the system.

Future power systems include a more active participation from the customer's side, thanks to the integration of demand-side management paradigms or the aggregation of various distributed generation and loads in the so-called virtual power plants.

2.3.1 Virtual power plants

Virtual power plants (VPPs) are a smart management paradigm consisting in coordinating generating units with different characteristics, stochastic or dispatchable ones, flexible profitable loads, and storage units in order to maximize the total profit [6]. Generation and consumption are managed in such a way that the market sees VPPs as a single flexible power output toward the grid.

The need for this new energy management paradigm is a consequence of the uncertainty and unreliability introduced by the diffusion renewable energy sources, in particular wind and solar power.

Introduction of a VPP leads to less risky and more efficient bidding in electricity markets, since bad predictions of renewable production can be adjusted by dispatchable generators, flexible loads, and energy storage. Furthermore, this paradigm allows the indirect access of renewable energy sources to the ancillary service markets, increasing their potential integration in power systems.

From the point of view of network operators, VPP simplifies system's management, since uncertainties and imbalances are locally addressed by the aggregator, which can assure a secure or at least less uncertain power production.

2.3.2 Demand response

Demand response refers to the active participation of customers to power system's balancing, requested by grid operators. Customers are encouraged to modify their load pattern by changes in price of electricity or incentive payments [7].

The transition toward smart grids is an essential element for the development of the demand response paradigm, since it requires advanced monitoring, communication, and control systems, in order to be correctly implemented in future power systems.

From the customer's point of view, the participation in a demand response program requires one of the following actions:

- *Load curtailment:* the customer simply reduces its load, accepting the consequences of this reduction.

- *Load shift:* the customer shifts the energy consumption to a different time interval.

- *Local generation:* the customer produces the energy needed locally, reducing its dependence from the grid.

Market operators can act on the demand response using different strategies that can be classified in three main branches:

- *Price programs:* customers pay time-varying prices during different periods of the day. The highest price is during the peak hours, while the lowest is set for off-peak hours.

- *Event-based programs:* customers are rewarded for changing their load upon request. Demand reduction signals may be sent to the participating customers that can voluntarily change their absorbed power in order to get the established incentive for a particular event.

- *Bid-based programs:* different customers send demand reduction bids, consisting in the amount of reduction capacity and the correspondent price. Depending on the bids, the cheaper ones are chosen.

Demand response has several benefits related to different domains. Customers participating in demand response programs have direct economical rewards;

however, all other customers benefit from the lower prices due to diminished demand peaks. Market performance is improved due to the lower market power of the producers, and operators have another flexible tool to address power systems' technical problems.

2.4 Operation domain

Operation domain is the set of actor and applications required for the reliable, safe, and efficient operation of power systems. The main duties are related to planning, monitoring, protection, maintenance, and control of power systems. The increasing diffusion of distributed renewable energy sources is posing a series of challenges to operators. Traditional power systems were designed for one-way power flows, while in modern grids, backward flows are frequent. Furthermore, the intermittent nature of solar and wind energy requires special attention to avoid system's imbalances and stability problems.

Smart grids are improving operators' capabilities with advanced technologies.

2.4.1 Wide area monitoring systems

Wide area monitoring systems (WAMS) are an emerging paradigm, involving the utilization of system-wide information to prevent the propagation of large disturbances, increasing the efficiency of the transmission system, and providing better protection and control [8]. Their development was possible thanks to the diffusion of advanced measurement systems: the phasor measurement units (PMU) that are the basic elements of WAMS.

The main difference between these sensors and the supervisory control and data acquisition (SCADA) systems is that they are all synchronized to a common reference time, and thus they can measure and compare the phase angles from voltage phasors of busses contained in a wide area control region. They allow a series of useful applications: state estimation, voltage stability control, dynamic thermal rating, congestion management, and fault localization.

The path toward a smarter grid implies the deployment of WAMS in power systems worldwide.

2.4.2 Smart meters

Smart meters are advanced energy meters that offer a wide range of functions, such as storage of detailed consumption data, two-way communication with utilities, and support for dynamic pricing, enabling demand response and monitoring of power quality and disturbance events. A proper deployment of such systems requires an automated metering infrastructure (AMI) that is a two-way communication network linking a huge number of smart meters.

Through AMIs, utilities can improve their analytic capabilities and operate systems in a more efficient, economic, and reliable way. Despite the benefits, AMIs could expose the system to a series of cyber security and customer's privacy threats; thus, they must be carefully designed in order to ensure the protection against cyberattacks.

2.5 Distribution domain

Distribution domain is electrically connected to the transmission domain and to the customer's domain. It is at lower voltage levels, and it is where most of the distributed generation has been installed during the last years.

Until the last decade, it was the least smart domain and the source of most of power outages and disturbances, due to the lack of monitoring and control capabilities. Smart grid transition is starting right from this domain, with a large deployment of smart systems, coupled with two-way communication links.

An evolving concept related to distributed generation, smart monitoring, and controls are microgrids.

2.5.1 Microgrids

Microgrids are smart interconnected networks of loads and distributed energy resources that can work connected to or separated from the electricity grid. They integrate a series of intelligent functions and pervasive controls, in order to be self-consistent [9].

They are characterized by six main components:

- *Power plants:* they can be dispatchable plants, such as low-power, fossil-fuel-based generators that recover waste heat, i.e., combined heat and power (CHP) systems or renewable generators, such as wind and solar plants.

- *Smart loads:* there can be residential, commercial, or industrial loads. In the case of microgrids, flexible and smart loads that can automatically control the absorbed power are preferred in order to better manage the whole system.

- *Energy storage:* storage systems are required to fully exploit renewable energy potential and to provide greater flexibility to the system.

- *Smart meters and sensors:* to keep the system working reliably, a huge amount of data related to different electrical quantities is needed for proper control.

- *Communication infrastructure:* to exchange information and commands between different elements.

- *Central management system:* data collected from microgrid's elements must be processed to take appropriate decisions by a central energy management system that sends command messages to all controllable devices.

Microgrids represent a gradual evolution from traditional grids to smart grids. Their control capabilities can ease grid operators' work, thanks to the local management of the system.

2.6 Service providers

Service providers deliver new and innovative services to producers, distributors, and customers. This domain is connected through communication flows to customer, operation, and market domains.

When delivering services, they preserve cybersecurity, reliability, stability, integrity, or safety of the electric power network.

Examples of these services are:

- *Customer management:* Managing relationships between customers and other power system's actors

- *Energy management:* Monitoring and controlling energy efficiency and demand of customers

- *Aggregation:* Aggregating different customers and producers and efficiently managing the system as a whole, enabling new paradigms such as virtual power plants

2.7 Customer domain

Customer domain is what defines the goal of a smart grid. This domain can be divided into three subdomains, each of which collects different customers with similar behaviors and energy needs: industrial, residential, and commercial. It is electrically connected to the distribution domain.

With the introduction of distributed energy sources, the customer is evolving to a *prosumer*, i.e., it both produces and consumes energy, and it has an active role in power systems.

Most of energy efficiency policies are addressed to this domain, and automation is playing a big role in reshaping it.

Key and innovative concepts for the customers of the future are demand response, described earlier, vehicle to grid (V2G) and energy hubs.

2.7.1 Vehicle to grid

Electrical vehicles are expecting to take an important place in vehicle market, due to the increasing focus on sustainability, energy supply security, and climate change. They have the potential to serve electric grid as dynamical energy storages. Since they are parked most of the time, they can remain connected to the grid [10].

Vehicle to grid is a concept that enables electrical vehicles to interact with power systems, in order to provide a series of functions to support the grid, such as peak power shaving, spinning reserve, voltage, and frequency regulation. Such opportunities are provided by electric vehicles through charging and discharging of their battery packs.

Their integration into power grids has to be carefully evaluated in order to avoid technical problems in distribution systems.

One of the most promising strategies for the integration of vehicle-to-grid technology is the aggregation of electrical vehicles to the virtual power plant or microgrid concepts, in such a way that energy flows are optimally controlled, taking into account power system's constraints, thus reducing eventual stress on the grid.

2.7.2 Energy hubs

A novel concept, potentially introducing further flexibility into power systems, is that of energy hubs.

An energy hub is a unit where multiple energy carriers, such as natural gas and electric energy, can be converted, conditioned, and stored. A typical example of converter is the CHP generator that produces both electric and thermal energy, taking natural gas as input [11].

The main advantages of energy hubs are the increased reliability of supply and higher flexibility. Such concept would add another choice to the participants to demand response, maintaining the same absorbed power but switching the primary energy source from electrical energy coming from the grid to another carrier, such as natural gas. In that case the customer would get the benefits of the demand response incentives without renouncing the part of the absorbed power.

Strictly connected to the concept of energy hubs is the energy interconnector, that is, the integrated transportation of electrical, chemical, and thermal energy, in a single underground device.

3. Conclusive remarks

The large-scale deployment of the smart grid technologies could lead existing power systems to evolve from static and hierarchical networks to decentralized and self-healing systems composed of cooperative and self-organizing energy resources. To this aim, in this chapter the most promising enabling technologies have been presented, and the possible research directions aimed at addressing some challenging open problems, which could hinder their deployment in existing power grids, have been outlined.

Author details

Alfredo Vaccaro[1*], Antonio Pepiciello[1] and Ahmed Faheem Zobaa[2]

1 Department of Engineering, University of Sannio, Benevento, Italy

2 Department of Electronic and Computer Engineering, Institute of Energy Futures, Smart Power Networks Theme, Brunel University London, UK

*Address all correspondence to: vaccaro@unisannio.it

References

[1] Farhangi H. The path of the smart grid. IEEE Power and Energy Magazine. 2010;**8**(1):18-28

[2] Ipakchi A, Albuyeh F. Grid of the future. IEEE Power and Energy Magazine. 2009;**7**(2):52-62

[3] Fang X, Misra S, Xue G, Yang D. Smart grid—The new and improved power grid: A survey. IEEE Communication Surveys and Tutorials. 2012;**14**(4):944-980

[4] ELicwPML. Office of the National Coordinator for Smart Grid Interoperability and I. T. Laboratory, Nist Framework and Roadmap for Smart Grid Interoperability Standards, release 2.0

[5] Mah D, Hills P, Li VO, Balme R. Smart Grid Applications and Developments. Green Energy and Technology: Springer; 2014

[6] Morales JM, Conejo AJ, Madsen H, Pinson P, Zugno M. Integrating Renewables in Electricity Markets: Operational Problems. Vol. 205. International Series in Operation Research and Management Science: Springer Science & Business Media; 2013

[7] Siano P. Demand response and smart grids—A survey. Renewable and Sustainable Energy Reviews. 2014;**30**: 461-478

[8] Terzija V, Valverde G, Cai D, Regulski P, Madani V, Fitch J, et al. Wide-area monitoring, protection, and control of future electric power networks. Proceedings of the IEEE. 2011;**99**(1):80-93

[9] Asmus P. Microgrids, virtual power plants and our distributed energy future. The Electricity Journal. 2010; **23**(10):72-82

[10] Mwasilu F, Justo JJ, Kim E-K, Do TD, Jung J-W. Electric vehicles and smart grid interaction: A review on vehicle to grid and renewable energy sources integration. Renewable and Sustainable Energy Reviews. 2014;**34**: 501-516

[11] Geidl M, Koeppel G, Favre-Perrod P, Klockl B, Andersson G, Frohlich K. Energy hubs for the future. IEEE Power and Energy Magazine. 2007;**5**(1):24-30

Chapter 2

Connected Autonomous Electric Vehicles as Enablers for Low-Carbon Future

Binod Vaidya and Hussein T. Mouftah

Abstract

Transportation is the main cause of various harmful gases being released into the atmosphere. Due to dependency on fossil fuels, conventional internal-combustion engine vehicles cause major impacts on air pollution and climate change. Achieving greenhouse gas (GHG) reduction targets requires electrification of transportation at the larger scale. Zero-emission vehicles are developing rapidly with consequences for energy use and GHG emissions, and their penetration is rising throughout the world. Such vehicles are widely considered as a promising solution for GHG reduction and a key to low-carbon mobility future. Recent trend in transportation system is a rapid shift toward connected autonomous vehicles. Connected autonomous electric vehicle (CAEV) will play a vital role in emerging revolution in sustainable low-carbon mobility. They can result in major reductions in GHG emissions and be at the forefront of rapid transformation in transportation. CAEVs have great potential to operate with higher vehicle efficiency, if they are charged using renewable energy sources that will significantly reduce emissions and dependency on fossil fuels. This book chapter is intended not only to provide understanding of potential environmental implications of CAEV technologies by reviewing the existing studies and research works but also to discuss environmental impacts including GHG emissions and improvement of vehicle efficiency.

Keywords: connected autonomous electric vehicle, CO_2 emissions, fuel economy, vehicle efficiency, low-carbon mobility

1. Introduction

Transportation is one of the leading causes of various harmful gases being released into the atmosphere. Particularly, due to dependency on fossil fuels, conventional Internal Combustion Engines (ICE) vehicles cause major impacts on air pollution and climate change along with negative impact on the well-beings of the society.

Achieving the global greenhouse gas (GHG) reduction targets requires the electrification of transportation at the larger scale. Zero-emission vehicles (ZEVs) that include hybrid electric vehicle (HEV), plugin hybrid electric vehicle (PHEV) and battery electric vehicles (BEV) are developing rapidly with consequences for energy use and GHG emissions. And the penetration of ZEVs is rising throughout the world. Though no country has yet accomplished a complete transition from conventional ICEs to ZEVs, some countries such as Norway and Netherlands are leading the way. Electric vehicles (EVs) are widely considered as a promising solution for GHG reduction and key to a low-carbon mobility future. It can be noted that BEV has the lowest amount of CO_2 emissions compared to ICE, HEV, and PHEV.

Recent trend in the transportation system is rapid shift toward autonomous vehicles (AV) [1]. A connected autonomous vehicle (CAV) is an emerging technology that could change the existing transportation system due to advanced communication and sensing capabilities, enhanced travel convenience, and the development of low-carbon mobility business models. Typically, the CAVs are electric, which are more efficient and therefore reduce the carbon emissions.

The connected autonomous electric vehicle (CAEV) will be an important part of the coming revolution in sustainable low-carbon mobility. Four major drivers that include automated driving, electric powertrains, connectivity, and shared mobility can provide compelling transition to a low-carbon future. Thus, they can result in major reductions in GHG emissions from transportation and are at the forefront of this rapid transformation in transportation [2–5]. These technological changes are outstretched and provide several opportunities and challenges.

The CAEVs have the great potential to operate with even higher vehicle efficiency, if they are charged using the electricity generated from renewable energy sources that will significantly reduce emissions as well as dependency on fossil fuels.

This book chapter is intended to provide understanding of the potential environmental implications of CAEV technologies by reviewing the existing studies and research works. We shall analyze environmental impacts including GHG emissions due to the transportation as well as improvement of vehicle efficiency.

2. Connected and autonomous electric vehicles

Connected and Autonomous Electric Vehicles (CAEVs) are complex automotive systems, combining basically connected vehicles (CV), autonomous vehicle (AV) and electric vehicle (EV) [1].

A connected vehicle (CV) is a vehicle with technology that enables it to communicate with nearby vehicles, infrastructure, as well as objects; but may not be automated nor electrically operated.

While, an autonomous vehicle (AV) is a vehicle that is, in the broadest sense, capable of driving itself without human intervention.

And electric vehicle (EV) is a vehicle that powers up and operates with energy stored in the battery.

Typically, CAEV is an electric vehicle that is capable of sensing its environment and navigating with little or no human input. CAEV senses its environment using various sensing devices including Radar, light detection and ranging (LiDAR), image sensors, 3D camera, etc. Basically, CAEV is composed of five major components.

- Perception system which is responsible for sensing the environment to understand its surroundings.

- Localization and mapping system that enables the vehicle to know its current location.

- Driving policy refers to the decision making capability of a CAEV under various situations, such as negotiating at roundabouts, giving way to vehicles and pedestrians, and overtaking vehicles.

- Communication system: as CAVs will be connected to the surrounding environment such as vehicles with vehicle to vehicle connectivity (V2V), to the infrastructure with vehicle to infrastructure (V2I) and to anything else such as the Internet: vehicle to anything (V2X), through wireless communications links.

- Storage battery system: this system includes charger and battery packs in the vehicle. Basically state of charge (SoC) level determines the amount of charge stored in the battery.

CAEVs definitely transform existing mobility paradigm. It can be observed that technological advancements in driving assistants and network connectivity yield further opportunities and services and meet the sustainable development for cleaner, safer, and smarter mobility. **Figure 1** shows CAEV applications and services.

2.1 Advantages of CAEVs

CAEVs offer many potential advantages in terms of sustainable development for environment friendly urban mobility, which are as follows [1].

- Improved safety: may eliminate many of the accidents caused by human error, estimated at about 90% of all accidents.

- Greater mobility: for those who cannot drive, including elderly, disabled, and youth.

- Reduced parking needs: passengers can be dropped off at their destinations without needing a nearby parking space.

Figure 1.
CAEV applications and services.

- Relaxed drivers: drivers can rest, work, or entertain themselves during a trip.
- Increased car-sharing: reduced need for individually-owned cars.
- Increased road capacity: through fleet platooning, more predictable traffic flow, and reduced congestion.
- Fewer CO_2 emissions and pollutants: using electric power to operate, can reduce GHG emissions as well as air pollution; minimized environmental impact; improve quality of life in urban area.
- Less fuel costs: fossil fuel will not be consumed to run CAEV, so fuel consumption is significant reduced.

With wide adoption of the CAEVs, it is expected to improve road safety, optimize traffic flow, help reduce fuel consumption, and minimize CO_2 emissions in the urban environments.

2.2 Classification of vehicle automation

Society of Automotive Engineers (SAE) released SAE International Standard J3016 that sets out taxonomy and standard to define different levels of autonomy. SAE updated its classification in 2016 as SAE J3016-201609.

Basically, vehicle automation has been categorized into various levels of autonomous vehicle technology ranging from Level 0, corresponding to no automation, to Level 5, corresponding to full automation. For instance, automated driver-assistance systems such as adaptive cruise control correspond to lower automation levels, while fully automated driverless vehicles correspond to higher automation levels.

The SAE defined levels of vehicle automation is depicted as follows:

Level 0—no automation:

In this level, the human driver is responsible for all the driving tasks including control of the car as well as monitoring the road and environment around the car.

Level 1—driver assistance:

In this level, the human driver is assisted with either steering or acceleration/deceleration by the driver assistance system but not both, for instance, adaptive cruise control.

Level 2—partial automation:

In this level, the driver assistance system take care of both acceleration/deceleration and steering control of the car, while the human driver monitors the road and environment around the car. It includes more advanced levels of driver assistance and requires continuous supervision of the driver.

Level 3—conditional automation:

In this level, the automated driving system undertakes all aspects of the dynamic driving task with the expectation that the human driver will respond appropriately to a request to intervene. Thus it requires partial supervision of the driver.

Level 4—high automation:

In this level, the automated driving system undertakes all aspects of the dynamic driving task, even if a human driver does not respond appropriately to a request to intervene. This level is basically unsupervised.

Level 5—full automation:

In this level, the automated driving system undertakes all aspects of the dynamic driving tasks in all roadway and environmental conditions. This level does not require driver at all.

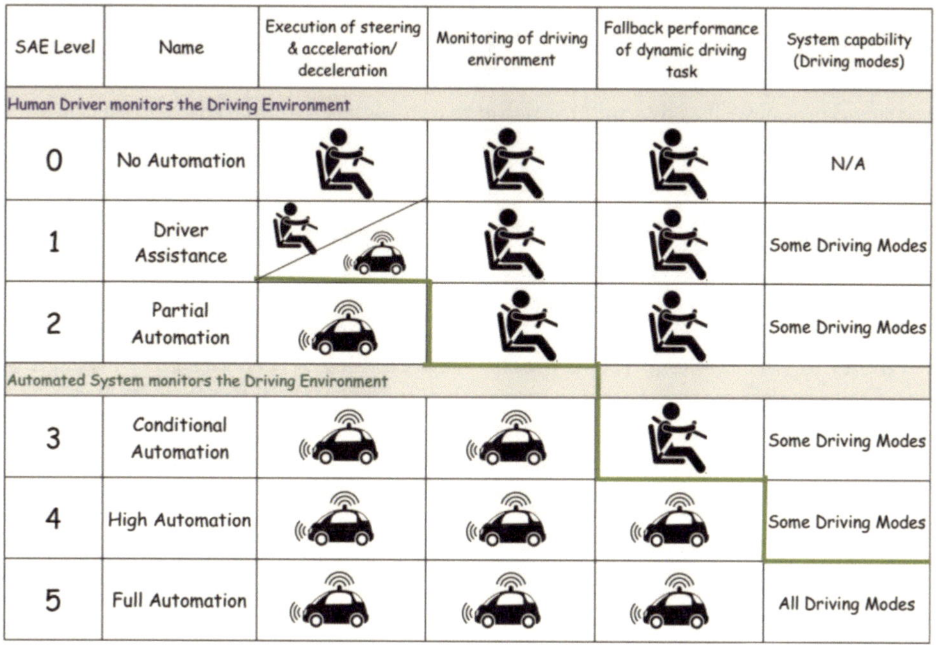

Figure 2.
The SAE-defined level of vehicle automation.

Figure 2 shows SAE defined level of vehicle automation.

3. CAEVs for low-carbon mobility

Mainly, due to four revolutions that include vehicular automation, vehicle electrification, vehicular connectivity, and shared mobility, CAEVs can offer great possibilities in expanding mobility and accessibility, and can play a leading role in achieving low-carbon mobility [6, 7].

These major drivers (i.e., automated driving, electric powertrains, connectivity, shared mobility) can provide compelling transition to a low-carbon future. In terms of energy use and carbon emissions, the potential synergies from combining these drivers would be significant.

The studies show that autonomous vehicles with electric powertrains have 40% lower lifetime GHG emissions than ICE-based vehicles. AV technologies along with V2V communications could smooth traffic flows and minimize braking, thus, possibly increasing fuel economy from 23 to 39% [8].

Similarly, the appealing on-demand shared mobility along with vehicular automation may offer the possibilities to expand multi-modal scenario that would reduce car travel by well over half in 2050, thus, would reduce traffic congestion as well as CO_2 emissions in 2050 less than one-third of the conventional vehicles [9].

3.1 Vehicular automation

As Vehicular automation involves the use of AI, and multi-agent system (MAS) to assist the vehicle operation, CAEV can be referred as a smart or intelligent.

Vehicular automation includes automated vehicle dynamics control such as adaptive cruise control (ACC) and automated powertrain operations that can improve vehicle energy efficiency and reduce carbon emissions.

Automated driving would allow to reduce unnecessary accelerations as well as decelerations so that energy efficiency can be improved. Consequently, Vehicular automation may result in optimized efficiency, increased safety, smooth traffic flow, few accidents as well as less pollution due to completely automated fuel control.

3.2 Vehicle electrification

Vehicle electrification, which is referred as zero emission, is one of the appealing ways to reduce transportation related fossil fuel consumption, in turn, minimize carbon emissions and other pollutions.

Due to the vehicle electrification, dependence on fossil fuel oil as well as vulnerability to volatile fuel prices is greatly reduced. Electric vehicles (EV) are significantly energy efficient than conventional ICE vehicles, thus, the formers have drastic reductions in long-term operation costs. Such an efficiency can be improved considerably if productive components (i.e., high-efficient motors, supercapacitors, high-efficient batteries) are utilized, electrical loss is reduced, and overall energy is optimized.

Furthermore, the electricity generation by hydro-power may not be quite clean, so cleaner electricity can be generated by using renewable energy sources such as photovoltaic (PV) solar, wind and other alternative energy sources such as hydrogen-fuel, bio-fuel. Using such technologies, vehicle electrification can provide significant impacts in energy efficiency.

3.3 Vehicular connectivity

Vehicular connectivity accommodates communication systems equipped within the vehicles that allow them to communicate with other vehicles and roadside units (RSU) to provide a wide range of information such as traffic, infotainment.

Motivation for vehicular communication systems is safety and reducing traffic collisions. Advancement of vehicular communication technologies (i.e., V2V, V2I) has not only revolutionized intelligent transportation system (ITS) but also furnishes various promising applications such as collision avoidance, dynamic traffic light.

Efficient use of the vehicular communications shall improve eco-driving (i.e., driving with efficiency maximizing speed, acceleration operating profiles and safety) and encourage more energy efficient driving, such as reducing traffic congestion and unnecessary stop-and-go operations at the intersections as well as shall optimize routing.

The vehicular connectivity can assist in enhancing multimode transportation that shall reduce VMT. The use of eco-driving techniques can improve fuel efficiency, thus in turn, reduce GHG emissions.

3.4 Shared mobility

Shared mobility is referred to Mobility-as-a-Service (MaaS), which can be described as a shift away from personally-owned modes of transportation toward mobility solutions that are consumed as a service. Shared mobility is evolving rapidly and can take several forms including ride-sharing, e-hailing, shared CAVs.

Due to enabling technologies including mobile and wireless, CAV in conjunction with shared mobility have the potential to increase the viability and shared transportation services.

The shared mobility services shall allow to increase roadway capacity by reducing number of vehicles on the road, thus reduce traffic congestion, and tailpipe emissions as well as reduce vehicle miles traveled (VMT), and vehicle energy consumption.

4. Environmental impacts of CAEVs

Adoption of CAEVs has the possibility to play a key role in addressing environmental impacts due to the existing transportation systems [10–13].

Deployment of CAEVs will yield immense changes in vehicle design, traveler behavior, mobility patterns and even urban planning that will have dramatic environmental impacts [14, 15].

Environmental impacts of CEAVs will depend on several factors. Some of them are listed as follows: effects of the total Vehicle-miles traveled (VMT); adoption of Mobility-as-a-Service (MaaS); Fuel economy; and eco-driving and platooning.

4.1 Vehicle miles traveled

Widespread deployment of CAEVs will lead to a dramatic rise in vehicle miles traveled (VMT) in future. The VMT will be increased mainly due to two factors: vehicles with empty occupancy before pick-up or after drop-off, and increasing demand of ride-hailing or ride-sharing services.

Zero occupancy vehicle travel might contribute significantly to VMT effects of CAEVs. Increase in VMT may lead to the increase in tailpipe emissions.

Adoption of CAEVs will provide independent mobility to non-drivers, including elderly people, people with disabilities, and youths. Since vehicle ownership among these groups will be very low, shared mobility services will be appealing among them. A study estimates an increment of 14% in the overall VMT as a result of wide penetration of CAEVs and mobility services provided to non-driving group.

A study also shows that personally-owned CAEVs are likely to significantly increase the total VMT and carbon emissions, as reduction in parking areas could exacerbate these increases by stimulating more zero occupancy vehicle travels.

4.2 Mobility-as-a-Service

The advancement of CAEV technology and the growth of on-demand shared mobility services may provide essential alternatives to conventional personally-owned vehicles, and have the potential to alter the way in which people move around cities.

Mobility-as-a-Service (MaaS) may be effective means to reduce VMT by combining trips that are temporally and spatially similar, and improve utilization of multi-modal transit providers as more users adopt MaaS as a main source of transit. Thus, MaaS may furnish several benefits including improved energy efficiency, traffic congestion reduction, and carbon emission reductions.

4.3 Fuel economy

Efficient driving generally furnishes increment in fuel economy. And an increase in fuel economy shall provide reduction in energy consumption, tailpipe emissions and air pollution [2].

More efficient driving in CAEVs can be achieved through a variety of mechanisms, including optimal driving cycle, dynamic eco-routing, traffic flow smoothing, and speed harmonization.

A wide-scale deployment of CAEVs could facilitate vehicle platooning that could lead to improved aerodynamics. These advances in CAEVs could lead to considerable improvements in fuel economy.

4.4 Eco-driving and platooning

Eco-driving may include route planning, trajectory optimization, and driving behavior improvement. And it is an effective way to reduce vehicle fuel consumption and achieve significant reduction in carbon emissions.

Platooning is based on Cooperative ACC (CACC) technologies that use V2V communication to enable constant time-gap following and ad hoc joining and leaving the platoon. Platooning dynamically chains CAEVs to maximize fuel efficiency.

Platooning is appealing due to the fact that it provides energy savings from aerodynamic drafting, more stable vehicle following dynamics, reduced traffic flow disturbances as well as potential safety improvements.

5. Discussion

Several research works, studies have been conducted for analyzing the environment impact of CAVs [16–18].

Collaborative research from Argonne National Laboratory, National Renewable Energy Laboratory (NREL), Oak Ridge National Laboratory, and U.S. Department of Energy illustrates the overall system VMT and fuel consumption effects due to vehicle automation [16].

In the study [16], the researchers have considered four scenarios: conventional vehicles, partial autonomy, full autonomy with ridesharing, and full autonomy without ridesharing, and associates upper (UB) and lower (LB) bounds with the latter two. **Figure 3** shows the upper bound and lower bound estimates on total U.S. Light-duty vehicle (LDV) fuel consumption for various CAV scenarios compared with the base Conventional scenario.

In **Figure 3**, it can be seen that there is large variation in the results; the total U.S. LDV fuel consumption (billion gallons per year) ranges from approximately 64% decrease in case of full autonomy with ridesharing LB up to approximately 205% increase in case of full autonomy without ridesharing UB.

Viewing the deployment of the CAVs (partial or full autonomy), the effects of travel demand and fuel efficiency were further investigated.

Possible changes in travel demand due to connectivity and increased vehicle automation are uncertain. The potential effects of CAVs on travel demand are classified into the following categories: less hunting for parking; easier travel; increased travel by under-served populations; mode shift from walking, transit and regional air; increase in ridesharing; and Increased empty miles traveled by automated vehicles [16].

Similarly, connectivity and vehicle automation have the potential to impact driving patterns, vehicle design as well as fuel efficiency. However, the impacts are uncertain. The categories of potential energy impact may include: drive profile and traffic flow smoothing; lesser congestion; efficient V2I/I2V communication; collision avoidance; platooning; and vehicle/powertrain resizing [16].

The researchers have elaborated the methodology that accommodates the above-mentioned travel demand and fuel consumption effect assumptions to estimate

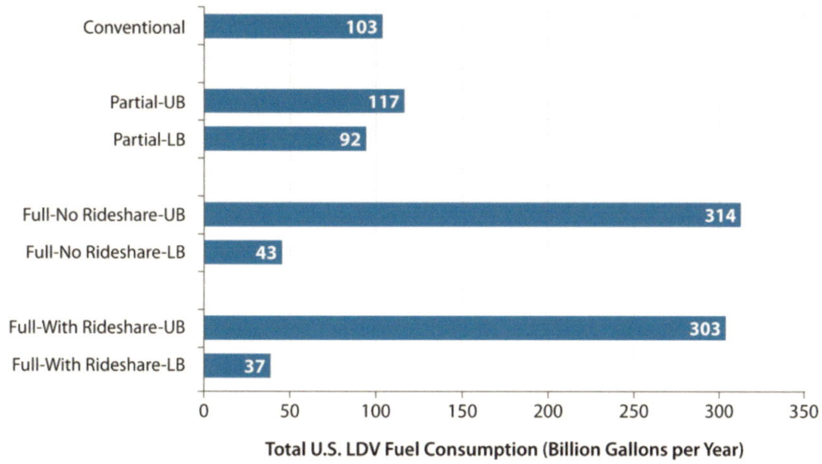

Figure 3.
Total U.S. LDV fuel consumption for various CAV scenarios compared with the base conventional scenario [16].

national-level fuel use impacts. **Table 1** shows the notations used in the equations for the computation of national fuel consumption impacts [16].

Referring to **Table 1**, the impacts $r_t^{i,j}$, $q_t^{i,j}$ are the fractional changes in the fuel consumption per mile over and above the fuel consumption per mile including all impacts considered earlier, that is, [16]

$$r_t^{i,j} = \left(\frac{FC_t^{i,j}}{FC_{t-1}^{i,j}}\right) - 1, q_t^{i,j} = \left(\frac{FC_t^{i,j}}{FC_{t-1}^{i,j}}\right) - 1 \qquad (1)$$

and analogously for $p_t^{i,j}$, and $s_t^{i,j}$:

$$p_t^{i,j} = \left(\frac{VMT_t^{i,j}}{VMT_{t-1}^{i,j}}\right) - 1, s_t^{i,j} = \left(\frac{VMT_t^{i,j}}{VMT_{t-1}^{i,j}}\right) - 1 \qquad (2)$$

Using the notations in **Table 1**, the baseline conventional fuel use in the U.S. (without CAVs) is calculated as: [16]

$$\sum\nolimits_{i\in I, j\in J}\left(VMT_0^{i,j} * FC_0^{i,j}\right) \qquad (3)$$

i	Set of road type I, {city, highways}
j	Set of time of day J, {peak hours, non-peak hours}
t	Set of technologies T, {partial automation technology, full automation technology}
$r_t^{i,j}$	Fuel impact estimated by partial automation technology t, on road type i, and time of day j
$q_t^{i,j}$	Fuel impact estimated by full automation technology t, on road type i, and time of day j
$p_t^{i,j}$	VMT impact estimated by partial automation technology t, on road type i, and time of day j
$s_t^{i,j}$	VMT impact estimated by full automation technology t, on road type i, and time of day j
$FC_0^{i,j}$	Original fuel consumption rate (gallon per mile), on road type i, and time of day j
$VMT_0^{i,j}$	Original vehicle miles traveled, on road type i, and time of day j

Table 1.
Notations used in equations for the computation of National fuel consumption impacts [16].

Consequently, the total fuel consumptions under partial automation and full automation scenarios can be calculated as follows: [16]

$$\sum_{i\in I, j\in J}\left(\left(VMT_0^{i,j}\prod_{t\in T}\left(1+p_t^{i,j}\right)\right)*\left(FC_0^{i,j}\prod_{t\in T}\left(1+r_t^{i,j}\right)\right)\right) \quad (4)$$

$$\sum_{i\in I, j\in J}\left(\left(VMT_0^{i,j}\prod_{t\in T}\left(1+s_t^{i,j}\right)\right)*\left(FC_0^{i,j}\prod_{t\in T}\left(1+q_t^{i,j}\right)\right)\right) \quad (5)$$

Based on the EPA Motor Vehicle Emission Simulator (MOVES) model values for U.S. national averages, the fraction of VMT on city and highway roads at peak and non-peak hours are estimated.

The analysis considers that the average fuel economy of LDVs to be 26.9 miles per gallon. Particularly, the analysis uses the relationship between city/highway fuel economy values with the combined fuel economy for the computation of the average city and highway fuel economy.

Additionally, it is assumed that traffic congestion occurs during peak hours and free flow driving occurs during non-peak hours. The analysis accounts the adjustment factors that have been suggested to calculate differences in fuel consumption (or GHG emission) under congestion and free flow driving and applies those adjustment factors in order to compute fuel economy and fuel consumption values during peak and non-peak hours. **Table 2** shows the assumptions of VMT in percent, fuel economy and fuel consumption for a baseline conventional vehicle under various road types and time of day [16].

The preceding formulations (Eqs. (1)–(5)), along with the assumptions from **Table 2**, yield the fuel consumption per mile impacts of various vehicle automation technologies at a national level, as shown in **Figure 4**. The effects are distinguished by partial and full automation CAVs [16].

In **Figure 4**, it can be seen that the adoption of full automation CAVs may have significant productive energy impacts. Typically, the increased fuel savings due to various categories are as follows: for vehicle/powertrain resizing, fuel saving is 0%–50%; for drive profile and traffic flow smoothing, it is 6.5%–16%; for platooning, it is 3%–5%; for intersection V2I/I2V communication, it is 2%–4%; and for collision avoidance, it is 0.2%–2.2%.

In the paper [17], the researchers have presented a methodological approach for refining this wide range of estimated fuel consumption.

The researchers have utilized a framework that accounts for energy impacts at the vehicle level, projected adoption levels, and changes in VMT in order to estimate national level fuel consumption impacts of CAVs. And they have considered several scenarios [17]

Road type/time of day	VMT %	Fuel economy (U.S. MPG)	Fuel consumption rate (U.S. GPM)
Highway, peak hours	18	29.7	0.0337
Highway, non-peak hours	27	35	0.0286
City, peak hours	22	21.4	0.0467
City, non-peak hours	33	25.2	0.0397

Table 2.
VMT percent, fuel economy, fuel consumption assumed for conventional vehicle by road type and time of day [16].

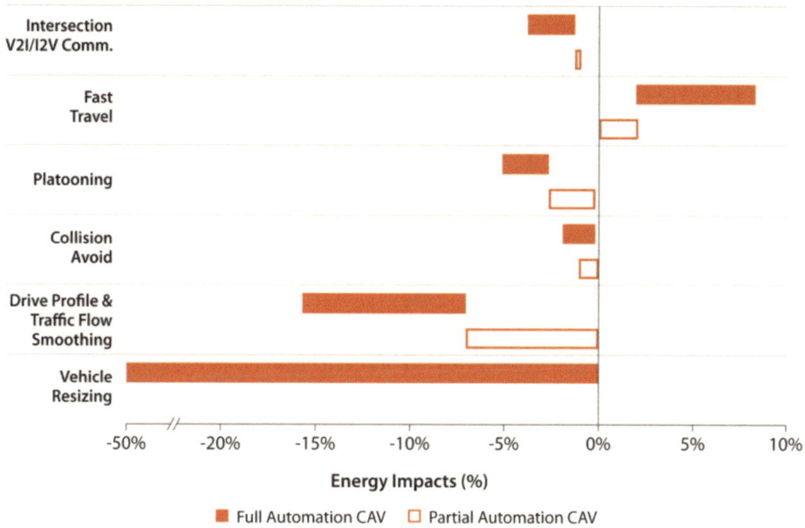

Figure 4.
Vehicle-level fuel consumption per mile impact under various vehicle automations [16].

- BASE-AEO is a scenario that is based on EIA's Annual Energy Outlook (AEO) 2017 Reference case;
- BASE-ADOPT is a scenario that is based on AEO 2017 inputs with projected vehicle sales shares from NREL's Automotive Deployment Options Projection Tool (ADOPT);
- CACC-AEO is a scenario with Cooperative Adaptive Cruise Control (CACC) penetration projections applied to the BASE-AEO case;
- CACC-ADOPT is a scenario with CACC applied to the BASE-ADOPT case;
- AutoTaxi-AEO is a scenario with automated taxis penetration projections applied to the BASE-AEO case; and
- AutoTaxi-ADOPT is a scenario with automated taxis applied to the BASE-ADOPT case.

Figure 5 shows U.S. total LDV fuel consumption for various scenarios (i.e., BASE-AEO, BASE-ADOPT, CACC-AEO, CACC-ADOPT, AutoTaxi-AEO, AutoTaxi-ADOPT) for a certain time period from 2015 to 2050 [17].

In **Figure 5**, it can be depicted that AutoTaxi scenerios (i.e., AutoTaxi-AEO, AutoTaxi-ADOPT) will have considerable energy impacts in compared with BASE use cases (i.e., BASE-AEO, BASE-ADOPT) in the future. For instance, by 2040, the US total LDV fuel consumption of CAVs decreases by 5% in case of AEO and 5.5% in case of ADOPT.

The paper [18] is based the well-established ASIF framework, which expresses transport carbon emissions in terms of the major drivers. The formulation for carbon emissions (E) can be stated in the following equation: [18].

$$E = A^{*} S^{*} I^{*} F \tag{6}$$

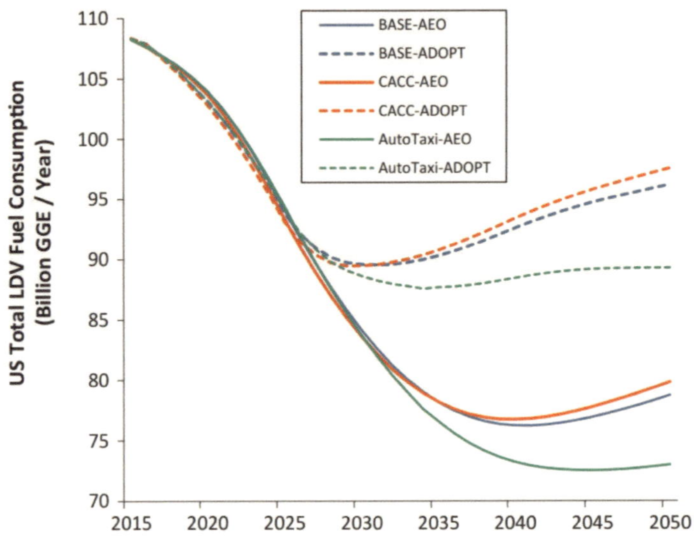

Figure 5.
U.S. total LDV fuel consumption scenarios for a certain time period from 2015 to 2050 [17].

where, A is activity level; S is modal share; I is energy intensity; and F is fuel carbon content.

The ASIF framework functions as a tool to organize various anticipated mechanisms through which vehicle automation may affect energy consumption and carbon emissions. Each driving factor on the right hand side of Eq. (6) can be considerably affected by the use of vehicular automation and thus fuel consumption and carbon emissions. **Table 3** illustrates a concise version of presumed mechanisms for energy impacts of automated vehicles (refer to [18] for details).

Figure 6 illustrates estimated ranges of possible energy impacts of vehicle automation in respect of various mechanisms [18].

Mechanism	ASIF element	Direction of effect	Automation level	Penetration level
Congestion mitigation	I	−ve	1–4	Moderate to high
Eco-driving	I	−ve	1–4	Any
Platooning	I	−ve	2–4	Any
Higher highway speeds	I	+ve	1–4	Moderate to high
De-emphasized performance	I	−ve	3,4	Any
Improved crash avoidance	I	−ve	2–4	Very high
Vehicle right-sizing	I	−ve	3,4	High to very high
Increased features	I	+ve	3,4	Any
Demand due to travel cost reduction	A,S	+ve	1–4	Any
Demand from New user groups	A,S	+ve	3,4	Any
Changed mobility services	A,S	−ve	3,4	Any
Potential for low-carbon transition	F	−ve	3,4	High

Table 3.
Presumed mechanisms for energy impacts of AVs [18].

It can be seen that several mechanisms may yield substantial reduction in energy use and carbon emissions, while others may have negative impacts. For instance, utilization of eco-driving, platooning, congestion mitigation, de-emphasized vehicle performance, lower crash risk, vehicle right-sizing, car-sharing and on-demand mobility, and reduced infrastructure footprint of automated vehicles may contribute to the improved energy efficiency of AVs. However, the increase in VMT due to lower travel costs, new user groups (youth, elderly, disabled), higher highway speeds, and increased vehicle features may increase the carbon footprints of AVs.

Table 4 provides abridged version of automation scenarios along with estimated ASIF multipliers for each effect. The scenarios vary in terms of levels of vehicle automation, effectiveness of the above-mentioned mechanisms in altering energy intensity, the degree of travel cost reductions, and the magnitude of travel demand [18].

The authors have conducted scenario analysis that shows vehicle automation may reduce energy use and GHG emissions by half in best-case scenario, or double them in a worst-case scenario, depending on the effects that come to dominate. Consequently, the outcomes depend on which scenarios prevail and proactive policy making is essential to steer the technology toward energy efficiency.

Overall energy and environmental implications of CAEVs in future will depend on following influencing mechanisms:

- Energy-saving algorithms and vehicle design.
- Vehicle operation (i.e., eco-driving).
- Platooning.
- Electrification using renewable energy resources.
- Changed mobility services.
- Vehicle utilization.
- Travel-cost implications.

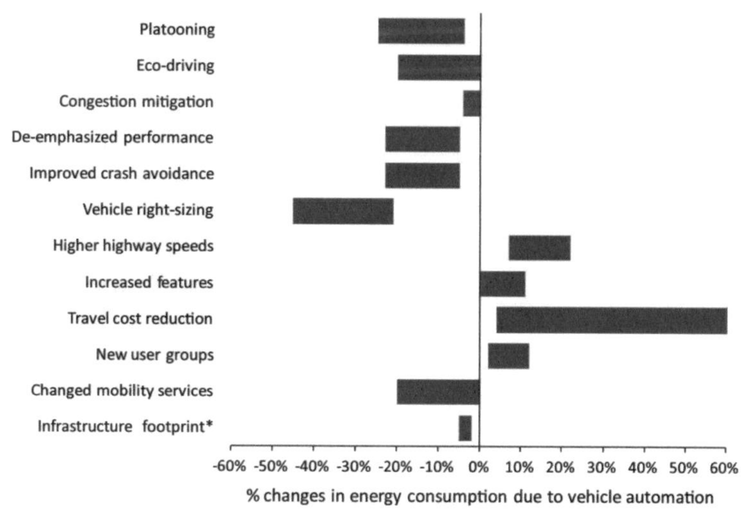

Figure 6.
Estimated ranges of energy impacts of vehicle automation in respect of various mechanisms [18].

	Mechanism	Scenario 1	Scenario 2	Scenario 3	Scenario 4
Energy intensity	Platooning	0.75	0.86	0.75	1.00
	Congestion	0.96	1.00	0.96	1.00
	Eco-driving	0.80	0.95	0.80	1.00
	Performance	0.77	1.00	0.77	1.00
	Crash avoidance	0.95	1.00	0.95	1.00
	Right-sizing	0.55	1.00	0.55	1.00
	Highway speeds	1.00	1.00	1.20	1.20
	Increased features	1.00	1.00	1.10	1.10
Travel demand	Generalized cost	1.56	1.09	1.89	1.49
	Newer user group	1.07	1.00	1.11	1.11
	Car-sharing	1.00	1.02	0.80	1.00

Table 4.
Automation scenarios along with estimated ASIF multipliers for each effect (abridged version) [18].

- Congestion mitigation.
- Information and communication technology (ICT) infrastructure.
- Government policies and laws.

Cybersecurity in CAEV networks is one of the active research areas [19–22]. Cybersecurity of CAEVs is essential for smart and sustainable development of a low-carbon city, since it may provide safety and social stability as well as economic sustainability.

Provisioning security and privacy in low-carbon smart mobility is crucial as without secure communications between CAEVs and remote systems may yield susceptibility to malicious attacks. For instance, compromised global position system (GPS) data affects the localization of CAEVs that may lead to traffic instability and/or hazardous accidents. Similarly, information shared among CAEVs in cooperative driving should be protected from any cyberattacks not only to guarantee road traffic safety but also to preserve privacy of CAEVs and other participating entities.

Current researches on CAEVs focus to identify cyber threats and vulnerabilities as well as to design strategies for preventing damages caused by these cyberattacks. Cyber threats and attacks studied include passive attacks such as eavesdropping, interception attack, traffic analysis and active attacks including impersonation attack, spoofing attack, replay attack, Sybil attack, jamming attack, message tampering [22]. Basically, requirements for cybersecurity solutions for CAEV networks may range from authentication, non-repudiation, integrity, to confidentiality.

Several open issues that should be addressed in future include [22]: in-vehicle security; security challenges in low-carbon smart cities; safety and security countermeasure consistency; and safe and secure mixed traffic systems.

6. Conclusion

The connected autonomous electric vehicle (CAEV) will be an important part of the coming revolution in sustainable low-carbon mobility. They can result in major reductions in GHG emissions from transportation and are at the forefront of this rapid transformation in transportation. The CAEVs have the great potential to operate with even higher vehicle efficiency, if they are charged using the electricity generated from renewable energy sources that will significantly reduce emissions as well as dependency on fossil fuels.

This book chapter provides the energy synergy of combining vehicular automation, vehicle electrification, and vehicular connectivity along with appealing on-demand mobility services. It also furnishes understanding of the potential environmental implications of CAEV technologies. Using several studies, the chapter highlights the analysis of environmental impacts including GHG emissions due to the transportation as well as improvement of vehicle efficiency.

Acknowledgements

This research work is supported by Smart Grid Fund (SGF), Ministry of Energy, The Ontario Government and Canada Research Chair (CRC) Fund, Canada.

Conflict of interest

The authors declare no conflict of interest.

Acronyms and abbreviations

Some of the useful acronyms and abbreviations used in this book chapter:

ICE	internal combustion engine
GHG	greenhouse gas
ZEV	zero-emission vehicle
AV	autonomous vehicle
CAV	connected and autonomous vehicle
CAEV	connected and autonomous electric vehicle
V2V	vehicle to vehicle
V2I	vehicle to infrastructure
VMT	vehicle mile traveled
MaaS	Mobility-as-a-Service

Author details

Binod Vaidya* and Hussein T. Mouftah
School of Electric Engineering and Computer Science, University of Ottawa, Ottawa, ON, Canada

*Address all correspondence to: bvaidya@uottawa.ca

References

[1] Siegel JE, Erb DC, Sarma SE. A survey of the connected vehicle landscape—Architectures, enabling technologies, applications, and development areas. IEEE Transactions on Intelligent Transportation Systems. 2018;**19**(8):2391-2406

[2] Hula A, Snapp L, Alson J, Simon K. The environmental potential of autonomous vehicles. In: Meyer G, Beiker S, editors. Road Vehicle Automation, Lecture Notes in Mobility. Vol. 4. Cham: Springer; 2018. pp. 89-95

[3] Alexander-Kearns M, Peterson M, Cassady A. The Impact of Vehicle Automation on Carbon Emissions: Where Uncertainty Lies. Center for American Progress; 2016. Available from: https://www.americanprogress.org/issues/green/reports/2016/11/18/292588/the-impact-of-vehicle-automation-on-carbon-emissions-where-uncertainty-lies/ [Accessed: 2018-10-20]

[4] MacKenzie D, Wadud Z, Leiby P. A first-order estimate of energy impacts of automated vehicles in the United States. In: Transportation Research Board 93rd Annual Meeting, TRB Paper No. 14-2193, Washington, DC, USA: 2014

[5] Dong J, Lu C, Hu L. Estimating Energy Efficiency of Connected and Autonomous Vehicles in a Mixed Fleet. Center for Transportation Research and Education, Iowa State University. Final Report; 2018

[6] Arbib J, Seba T. Rethinking Transportation 2020–2030: The Disruption of Transportation and the Collapse of the Internal-Combustion Vehicle and Oil Industries. RethinkX: Disruptions, Implications and Choices; 2017

[7] Mersky AC, Samaras C. Fuel economy testing of autonomous vehicles. Transportation Research Part C. 2016;**65**:31-48

[8] Vahidia A, Sciarretta A. Energy saving potentials of connected and automated vehicles. Transportation Research Part C. 2018;**95**:822-843

[9] Brown A, Gonder J, Repac B. An analysis of possible energy impacts of automated vehicles. In: Meyer G, Beiker S, editors. Road Vehicle Automation, Lecture Notes in Mobility. Cham: Springer; 2014. pp. 137-153

[10] Rubin J. Connected autonomous vehicles: Travel behavior and energy use. In: Meyer G, Beiker S, editors. Road Vehicle Automation, Lecture Notes in Mobility. Vol. 3. Cham: Springer; 2016. pp. 151-162

[11] Auld J, Verbas O, Javanmardi M, Rousseau A. Impact of privately-owned level 4 CAV technologies on travel demand and energy. Procedia Computer Science. 2018;**130**:914-919

[12] Eilbert A, Noel G, Jackson L, Sherriff I, Smith S. Potential Energy and Emission Benefits of Vehicle Automation and Connectivity. U.S. Department of Transportation: Volpe National Transportation Systems Center; 2017

[13] Qi X, Barth MJ, Wu G, Boriboonsomsin K, Wang P. Energy impact of connected eco-driving on electric vehicles. In: Meyer G, Beiker S, editors. Road Vehicle Automation. Vol. 4. Cham: Lecture Notes in Mobility, Springer; 2018. pp. 97-111

[14] Chen Y, Young S, Qi X, Gonder J. A first-order estimate of automated mobility district fuel consumption and GHG emission impacts. In: Meyer G, Beiker S, editors. Road Vehicle Automation, Lecture Notes in Mobility.

Vol. 4. Cham: Springer; 2018. pp. 113-123

[15] Rodier C. Travel Effects and Associated Greenhouse Gas Emissions of Automated Vehicles. White paper from The National Center for Sustainable Transportation, Institute for Transportation Studies, UC Davis; 2018

[16] Stephens TS, Gonder J, Chen Y, Lin Z, Liu C, Gohlke D. Estimated bounds and important factors for fuel use and consumer costs of connected and automated vehicles. Technical Report NREL/TP-5400-67216. National Renewable Energy Laboratory, CO, USA; 2016

[17] Stephens TS, Auld J, Chen Y, Gonder J, Kontou E, Lin Z, et al. Assessing energy impacts of connected and automated vehicles at the U.S. National Level—Preliminary bounds and proposed methods. In: Meyer G, Beiker S, editors. Road Vehicle Automation, Lecture Notes in Mobility. Vol. 5. Cham: Springer; 2019. pp. 105-115

[18] Wadud Z, MacKenzie D, Leiby P. Help or hindrance? The travel, energy and carbon impact of highly automated vehicles. Transportation Research Part A. 2016;**86**:1-18

[19] Petit J, Shladover SE. Potential cyberattacks on automated vehicles. IEEE Transactions on Intelligent Transportation Systems. 2015;**16**(2): 546-556

[20] Parkinson S, Ward P, Wilson K, Miller M. Cyber threats facing autonomous and connected vehicles: Future challenges. IEEE Transactions on Intelligent Transportation Systems. 2017;**18**(11):2898-2915

[21] Karnouskos S, Kerschbaum F. Privacy and integrity considerations in hyperconnected autonomous vehicles. Proceedings of the IEEE. 2018;**106**(1): 160-170

[22] Cui J, Liew LS, Sabaliauskaite G, Zhou F. A review on safety failures, security attacks, and available countermeasures for autonomous vehicles. Ad Hoc Networks. 2018. DOI: 10.1016/j.adhoc.2018.12.006

Chapter 3

A Distributed Optimization Method for Optimal Energy Management in Smart Grid

Dinh Hoa Nguyen, Huynh Ngoc Tran, Tatsuo Narikiyo and Michihiro Kawanishi

Abstract

This chapter presents a distributed optimization method named sequential distributed consensus-based ADMM for solving nonlinear constrained convex optimization problems arising in smart grids in order to derive optimal energy management strategies. To develop such distributed optimization method, multi-agent system and consensus theory are employed. Next, two smart grid problems are investigated and solved by the proposed distributed algorithm. The first problem is called the dynamic social welfare maximization problem where the objective is to simultaneously minimize the generation costs of conventional power plants and maximize the satisfaction of consumers. In this case, there are renewable energy sources connected to the grid, but energy storage systems are not considered. On the other hand, in the second problem, plug-in electric vehicles are served as energy storage systems, and their charging or discharging profiles are optimized to minimize the overall system operation cost. It is then shown that the proposed distributed optimization algorithm gives an efficient way of energy management for both problems above. Simulation results are provided to illustrate the proposed theoretical approach.

Keywords: optimal energy management, distributed optimization, social welfare maximization, ADMM, smart grid, multi-agent system, consensus, electric vehicle

1. Introduction

In any energy system, optimal energy management (OEM) is an essential problem because it directly affects to both the technical (e.g., operation and control) and economic (i.e., profit) aspects of such energy system. Recently, smart grid has been proposed as a solution to improve the greenhouse gas emissions and the efficiency and energy management in electric power grids [1]. Important components in smart grid are renewable energy sources (RESs) and distributed energy resources (DERs), e.g., rooftop photovoltaic (PV), electric vehicles (EVs), distributed energy storage systems (ESSs), etc. Those RESs and DERs are expected to replace the polluted fossil-based energy sources for generating electric power and to increase the mobility and flexibility of power grids.

However, the fluctuating and intermittent nature of RESs and the close or on-site location of DERs to end users make the OEM problem in smart grids more complex than that in traditional power grids. This urges the development of new concepts and approaches to deal with challenges that arose in smart grids. Some examples include demand side management (DSM) and real-time (dynamic) pricing (RTP) [2, 3]. As one of the main DSM activities [3–5], demand response (DR) motivates changes in electric use by end-use customers, in response to changes in the price of electricity over time, or gives incentive payments designed to induce lower electricity use at times of high market prices or when grid reliability is jeopardized [6]. DR not only shifts the load in peak times but also increases the grid's energy efficiency and customers' benefits.

Traditionally, the OEM problem in transmission power networks is treated as the economic dispatch (ED) problem (see, e.g., [7]). Nonetheless, DSM including DR is often not included in the ED problem. To overcome this drawback, the social welfare maximization (SWM) problem is set up and solved, where the benefits from both the suppliers and the consumers are taken into account [8–14]. The effectiveness of the SWM problem has been verified in [15]. On the other hand, in distribution power networks, the OEM problem is usually investigated with the existence of an intermediate player called the aggregator who manages the power exchange between a small distribution grid and a larger or main grid. Therefore, the OEM problem in such situations means to maximize the benefits of the DER owners, consumers, and the aggregator.

To suppress the redundancy or to supplement the lack of energy from the fluctuating outputs of RESs, and to make the energy scheduling more flexible, ESSs are being introduced into smart grids as a solution. For instance, in power distribution systems or microgrids, EV's batteries are sources of mobile and distributed ESSs that might contribute to the energy efficiency, security, and flexibility of such distribution systems or microgrids. The stability and robustness of vehicle-to-grid (V2G) energy networks have been investigated in [16]. The problem of power scheduling with electric vehicle (PSwEV) in a microgrid has been introduced under a multi-agent setting in [17]. Due to the increase of market share of EVs and plug-in hybrid electric vehicles (PHEVs) predicted until 2050 [18], it is reasonable to believe that the PSwEV problem will be popular and important in the near future.

To solve optimization problems arising in OEM problems, the existing methods can be classified depending on the centralized or distributed nature of such methods and the heuristic or analytical characteristic of the obtained solution. The security and resiliency of centralized approaches are weak due to a single point of failure and the huge communication and data processing, at the central unit. Those limitations can be improved by distributed approaches using multi-agent system (MAS), where each agent corresponds to a bus or a portion of the grid and each agent only communicates with a few other neighboring agents. On the other hand, the heuristic methods (e.g., [19, 20]) usually require a long running time and local solutions, which are less effective than analytical methods for convex optimization problems. Thus, distributed and analytical approaches will be developed in the current chapter to solve convex optimization problems representing OEM problems in smart grids.

A MAS-based distributed method for solving the ED problem in smart grid was proposed in [7] where the power losses are approximated by quadratic functions and the nonlinear coupling of oscillating agents is employed for decentralized solution derivation from the Karush-Kuhn-Tucker (KKT) conditions. The projected gradient methods were utilized in [10] to solve the SWM problem, where a MAS was utilized to derive the supply-demand mismatch in a distributed fashion. Another method named dual decomposition was used in [8] to get a distributed solution when the power balance is not strictly required. If the power balance is required, a distributed observer design was employed in [11] to estimate the power mismatch. The Vickrey-Clarke-Groves (VCG) mechanism was used in [9], but it is centralized. A distributed method was presented in [12], but the communication structure was all-to-all. Next, two consensus protocols were introduced in [13] to derive a distributed method for solving the SWM problem. Lastly, the SWM problem with transmission losses modeled by a quadratic function was investigated in [14].

This chapter presents an approach named sequential distributed consensus-based ADMM (SDC-ADMM) for solving nonlinear convex optimization problems having both equality and inequality constraints, which include those from OEM problems in smart grids. The attractive features of this approach are as follows: (i) viable for a general and broad class of constrained convex programming, (ii) distributed implementation, (iii) analytical updates of variables, and (iv) unnecessary checking of the active constraints in the KKT conditions, e.g., as in [7], which slows down the convergence speed and increases the computational complexity. Then the effectiveness of the proposed approach will be demonstrated through two specific OEM problems in smart grids, namely, dynamic SWM (DSWM) and PSwEV.

The rest of this chapter is organized as follows. Section 2 introduces a class of constrained optimization problems in power grids followed by two specific problems. Then an approach to solve those optimization problems is given in Section 3. Simulation results to illustrate the proposed approach are provided in Section 4. Finally, Section 5 summarizes the chapter.

2. Constrained optimization problems in smart grid

A lot of OEM problems in power grids can be represented by the following general form of constrained convex programming:

$$\min \sum_{t=1}^{N} \sum_{i=1}^{M} f_i(x_i(t)) \tag{1}$$

$$\text{s.t.} \sum_{i=1}^{M} \mu_i x_i(t) = \xi(t) \tag{2}$$

$$x_i(t) \in \Omega_i(t) \tag{3}$$

where $f_i : \mathbb{R} \to \mathbb{R}$ are scalar, continuous, and convex functions, $x_i \in \mathbb{R}$ are variables, t is the time index, $\xi(t)$ is a known time-varying parameter, $\mu_i > 0$ are known parameters, $\Omega_i(t) \subset \mathbb{R}$ are local convex constraint sets which might be dependent on the previous time slot $t-1$, $x(t) \triangleq [x_1(t), \ldots, x_M(t)]^{\mathrm{T}}$, and N and M are the number of time slots and variables.

Due to the global constraint (2), (1) is a centralized optimization problem. However, (1) is decomposable into subproblems corresponding to individual time slots; hence we can derive a sub-optimal solution to (1) by resolving it sequentially, i.e., from $t = 1$ to $t = N$. Doing so saves much time and computational costs in comparison with solving (1) for all time slots at once. Therefore, a sequential approach to solve (1) will be proposed in this chapter and will be applied to two problems, namely, DSWM and PSwEV. The former aims to maximize the benefits of both generators and consumers in transmission grids without explicitly considering ESSs, while the latter investigates the power scheduling in microgrids with EVs serving as mobile and distributed ESSs.

2.1 Dynamic social welfare maximization problem

Consider a smart grid consisting of n conventional generation (CG) units, q renewable generation (RG) units, and m demand units. Denote $P_{1,G}(t), \ldots, P_{n,G}(t)$ the powers generated by CG units; $P_{1,R}(t), \ldots, P_{q,R}(t)$ the powers generated by RG units; and $P_{1,D}(t), \ldots, P_{m,D}(t)$ the powers consumed by demand units; and $P(t) \triangleq [P_{1,G}(t), \ldots, P_{n,G}(t), P_{1,D}(t), \ldots, P_{m,D}(t)]^T; P \triangleq \left[P(1)^T, \ldots, P(N)^T\right]^T$.

The generation cost for CG unit i is $C_i(P_{i,G}(t)) = a_i P_{i,G}^2(t) + b_i P_{i,G}(t) + c_i$, with prescribed coefficients a_i, b_i, c_i. The utility function for demand unit j [8–10], which shows the consumer's satisfaction with respect to its consumed power, is

$$U_j(P_{j,D}(t)) = \begin{cases} \beta_j P_{j,D}(t) - \alpha_j P_{j,D}^2(t) & : P_{j,D}(t) \leq \dfrac{\beta_j}{2\alpha_j}, \\ \dfrac{\beta_j^2}{4\alpha_j} & : P_{j,D}(t) \geq \dfrac{\beta_j}{2\alpha_j} \end{cases}$$

where α_j and β_j are predetermined parameters. Then the DSWM problem in smart transmission grids is as follows:

$$\min_{\mathbf{P}} \sum_{t=1}^{N} \left[\sum_{i=1}^{n} C_i(P_{i,G}(t)) - \sum_{j=1}^{m} U_j(P_{j,D}(t)) \right] \quad (4)$$

$$\text{s.t.} \sum_{i=1}^{n} P_{i,G}(t) + \sum_{l=1}^{q} P_{l,R}(t) = \sum_{j=1}^{m} P_{j,D}(t) + P_L(t) \quad (5)$$

$$P_{i,G}^{\min} \leq P_{i,G}(t) \leq P_{i,G}^{\max} \quad (6)$$

$$\Delta P_{i,G}^{\min} \leq P_{i,G}(t) - P_{i,G}(t-1) \leq \Delta P_{i,G}^{\max} \quad (7)$$

$$P_{j,D}^{\min} \leq P_{j,D}(t) \leq P_{j,D}^{\max} \quad (8)$$

where $P_L(t)$ is the total power loss in the grid at time slot t, (5) is the power balance constraint, (6)–(7) represent the realistic limits on the output powers of CG units and their ramp rates, and (8) is a constraint for consumed powers of demand units. Next, denote the power loss coefficients of CG unit i, demand unit j, and RG unit l by $\gamma_{i,G} \triangleq \frac{\partial P_L}{\partial P_{i,G}}, \gamma_{j,D} \triangleq \frac{\partial P_L}{\partial P_{j,D}}, \gamma_{l,R} \triangleq \frac{\partial P_L}{\partial P_{l,R}}$, respectively [10]. Then the power balance constraint (5) can be rewritten as

$$\sum_{i=1}^{n} (1 - \gamma_{i,G}) P_{i,G}(t) + P_R(t) = \sum_{j=1}^{m} \left(1 + \gamma_{j,D}\right) P_{j,D}(t)$$

where $P_R(t) \triangleq \sum_{l=1}^{q} (1 - \gamma_{l,R}) P_{l,R}(t)$. On the other hand, for $i = 1, \ldots, n$, (6) and (7) can be conveniently rewritten as

$$P_{i,G}(t) \in \Omega_i(t) \triangleq \left\{ x_i(t) : \hat{P}_{i,G}^{\min,t} \leq x_i(t) \leq \hat{P}_{i,G}^{\max,t} \right\}$$

where

$$\hat{P}_{i,G}^{\min,t} \triangleq \max\left\{P_{i,G}^{\min}, \Delta P_{i,G}^{\min} + P_{i,G}(t-1)\right\}, \hat{P}_{i,G}^{\max,t} \triangleq \min\left\{P_{i,G}^{\max}, \Delta P_{i,G}^{\max} + P_{i,G}(t-1)\right\}$$

for $t = 2, \dots, N$, and $\hat{P}_{i,G}^{\min,t} \triangleq P_{i,G}^{\min}$, $\hat{P}_{i,G}^{\max,t} \triangleq P_{i,G}^{\max}$ for $t = 1$. In addition, (8) is rewritten as

$$P_{j,D}(t) \in \Omega_j(t) \triangleq \left\{x_j(t) : P_{j,D}^{\min} \le x_j(t) \le P_{j,D}^{\max}\right\}, j = n+1, \dots, n+m$$

2.2 Power scheduling with electric vehicle

In this problem, we include battery operation into the microgrid to suppress the high demand at the high-cost time of utility electricity. Furthermore, we consider EV battery instead of the stationary battery storage to reduce the installation cost of the battery system.

Our microgrid model consists of n diesel generations (DGs), m demand units, q PV generations, v EVs, and *one* microgrid operator. This microgrid is connected to the main grid (MG) whose electricity trading price is predescribed as $q(t)$. The electricity trading price inside the microgrid, one of decision variables, is denoted by $p(t)$.

2.2.1 Diesel generation and load demand

The generation cost function $C_i(P_{i,G}(t))$ for DG unit i is similar to that in the Section 2.1. The formulas of constraints for DG powers are also the same to that shown in (6) and (7), i.e.,

$$P_{i,G}^{\min} \le P_{i,G}(t) \le P_{i,G}^{\max} \tag{9}$$

$$\Delta P_{i,G}^{\min} \le P_{i,G}(t) - P_{i,G}(t-1) \le \Delta P_{i,G}^{\max} \tag{10}$$

For the microgrid internal trading, the revenue function of DGs has the following form:

$$W_{i,G}(P_{i,G}(t), p(t)) = p(t)P_{i,G}(t) - C_i(P_{i,G}(t)) \tag{11}$$

Load demands are assumed as fixed parameters in this problem. The electricity cost of load demands is calculated by

$$W_{j,D}(P_{j,D}(t), p(t)) = p(t)P_{j,D}(t) \tag{12}$$

2.2.2 Electric vehicle

For simplicity, we assume that EVs have only one round-trip route per day, and the home of each EV owner is the only charging point for each EV.

Denote $P_{h,EV}(t)$ the charging/discharging power of the *h-th* EV during time slot t ($P_{h,EV}(t) > 0$ in charging mode and $P_{h,EV}(t) < 0$ in discharging mode), $P_{h,EV}(t)$ need to satisfy the following constraint:

$$P_{h,EV}^{min}(t) \le P_{h,EV}(t) \le P_{h,EV}^{max}(t) \tag{13}$$

To reduce computational cost, $P_{h,EV}(t)$ between departure and arrival time slots of a route are assumed by zero, without loss of generality, and $P_{h,EV}(t)$ at these departure and arrival time slots are assumed to be equal to a half of $-P_h^{drive}$, where

P_h^{drive} is the equivalently consumed power of the *h-th* EV as if the EV route just lasts for one time slot.

$$P_{h,EV}\left(T_h^{dep} < t < T_h^{arr}\right) = 0 \tag{14}$$

$$P_{h,EV}\left(T_h^{dep}\right) = P_{h,EV}\left(T_h^{arr}\right) = -0.5P_h^{drive} \tag{15}$$

The state of charge (SOC) of EV battery at the starting point of a next time slot depends on the SOC at the starting point of current time slot and a charging/discharging efficiency φ_h, as shown in (4).

$$SOC_h(t+1) = \begin{cases} SOC_h(t) + \varphi_h P_{h,EV}(t)\Delta t & : P_{h,EV}(t) \geq 0 \\ SOC_h(t) + \dfrac{P_{h,EV}(t)\Delta t}{\varphi_h} & : P_{h,EV}(t) < 0 \end{cases} \tag{16}$$

SOC constraint is as follows:

$$SOC_h^{\min} \leq SOC_h(t) \leq SOC_h^{\max} \tag{17}$$

When EVs are at home (from the arrival time to the next departure time), their charging/discharging scheduling can be utilized for DR actions. The variable $P_{h,EV}(t)$ during EV plugged-in time is the decision variable and will be solved by the proposed algorithm. However, it is necessary to ensure that EVs have enough energy (SOC) for their routes before their departure times. To satisfy this requirement, we consider a simple EV charging strategy based on [21]–[22] in which $P_{h,EV}^{min}$ and $P_{h,EV}^{max}$ in (13) are specified during the EV plugged-in time. The EV charging strategy is presented below.

Algorithm 1: EV charging strategy (from T_h^{arr} to the next day T_h^{dep})

1: **Input:** Departure time T_h^{dep}, and the required state-of-charge SOC_h^{req}
2: $t_0 = round\left(T_h^{dep} - \frac{SOC_h^{req}}{P_{EV}^{max}\Delta t}\right)$
3: **If** $T_h^{arr} \leq t < t_0$
4: $P_{h,EV}^{max}(t) = -P_{h,EV}^{min}(t) = P_{EV}^{max} = -P_{EV}^{min} = const$
5: $t = t + 1$
6: **Return** to 3
7: **Else** $t_0 \leq t < T_h^{dep}$
8: $P_{h,EV}^{min}(t) = max\left(P_{EV}^{min}, P_{EV}^{max}(t - t_0) - \left(SOC_h(t) - SOC_h^{min}\right)/\Delta t\right)$
9: $P_{h,EV}^{max}(t) = min\left(P_{EV}^{max}, \left(SOC_h^{max} - SOC_h(t)\right)/\Delta t\right)$
10: $t = t + 1$
11: **Return** to 3
12: **Else** $t = T_h^{dep}$
13: **End**

The battery life function of EVs depends on EVs charging/discharging power [23], as follows:

$$L_h(P_{h,EV}(t)) = -\mu_h\, P_{h,EV}^2(t) + \pi_h P_{h,EV}(t) P_{h,EV}(t-1) \tag{18}$$

where μ_h and π_h are constant coefficients, $h = 1, \ldots, v$.

The revenue function of EVs is defined in [21], taking into account the battery life cost:

$$W_{h,EV}(P_{h,EV}(t),p(t)) = L_h(P_{h,EV}(t)) - p(t)P_{h,EV}(t) \tag{19}$$

2.2.3 Microgrid operator

The revenue function of the microgrid operator is calculated by

$$W_{mg}(P_g(t),p(t)) = (p(t) - q(t))P_g(t) \tag{20}$$

where $P_g(t)$ is the power trading between the MG and the microgrid via the microgrid operator. In the case that the microgrid sells power to the MG, $P_g(t)$ is negative. Furthermore, $P_g(t)$ needs to satisfy upper and low bounds, as follows:

$$P_g^{\min} \le P_g(t) \le P_g^{max} \tag{21}$$

2.2.4 Power balance constraint in the microgrid

With the existence of EVs and the MG, the power balance constraint in the microgrid has more terms than that in the DSWM problem, as follows:

$$\sum_{i=1}^{n} P_{i,G}(t) + \sum_{l=1}^{q} P_{l,R}(t) + P_g(t) = \sum_{j=1}^{m} P_{j,D}(t) + \sum_{h=1}^{v} P_{h,EV}(t) \tag{22}$$

2.2.5 PSwEV optimization problem

The PSwEV problem is to maximize the total revenue of DGs, EVs, load demands, and the microgrid operator, as follows:

$$\max_{\mathbf{P}} \sum_{t=1}^{N} g(P(t)) \tag{23}$$

s.t. (22), (9), (10), (13)–(19), and (21)
where

$$P(t) \triangleq \left[P_{1,G}(t), \ldots, P_{n,G}(t), P_{1,EV}(t), \ldots, P_{v,EV}(t), P_g(t)\right]^T; \mathbf{P} \triangleq \left[P(1)^T, \ldots, P(N)^T\right]^T$$

$$g(P(t)) \triangleq \sum_{i=1}^{n} W_{i,G}(P_{i,G}(t),p(t)) - \sum_{j=1}^{m} W_{j,D}(p(t)) + \sum_{h=1}^{v} W_{h,EV}(P_{h,EV}(t),p(t)) + W_{mg}(P_g(t),p(t))$$

Substituting (11), (12) and (18)–(20) into (23), the PSwEV maximization programming in (23) can be rewritten as a minimization problem below:

$$\min_{\mathbf{P}} \sum_{t=1}^{N} \left(\sum_{i=1}^{n} C_i(P_{i,G}(t)) - \sum_{h=1}^{v} L_h(P_{h,EV}(t)) + q(t)P_g(t) \right) \tag{24}$$

s.t. (22), (9), (10), (13)–(19), and (21)

3. Sequential distributed consensus-based ADMM approach

As mentioned in Section 2, the optimization problem (1) will be solved by a sequential approach presented in the current section, which separates (1) into the following subproblems

$$\min \sum_{i=1}^{M} f_i(x_i(t)) \tag{25}$$

$$\text{s.t.} \sum_{i=1}^{M} \mu_i x_i(t) = \xi(t) \tag{26}$$

$$x_i(t) \in \Omega_i(t) \tag{27}$$

and solves them starting from $t = 1$ until $t = N$. Thus, (25) is well-defined and the time index t can be dropped for conciseness. Then to solve (25), existing methods can be utilized, e.g., gradient-based methods, dual decomposition method, etc. Nevertheless, an approach called sequential distributed consensus-based ADMM (SDC-ADMM), which combines the advantages of the aforementioned methods [24] and avoids problems of centralized approaches, will be proposed in this section.

3.1 Multi-agent system description for smart grid

The SDC-ADMM approach is based on MAS and consensus theory so that it can be run in parallel in all generation and consumption units. Hence, the MAS description for smart grid needs to be introduced first. More specifically, each agent is assigned to a generation or demand unit, and the communication among agents is represented by an undirected graph $\mathcal{G}$. Each node represents a unit in the grid whose variable is x_i or ξ, and each edge represents the communication between two nodes. For each node i, denote $\aleph_i$ its neighbor set and $|\aleph_i|$ the cardinality of $\aleph_i$.

3.2 Reformulation of smart grid optimization problems

To develop the SDC-ADMM approach, (25) is first reformulated into the 2-block form of ADMM. The following closed convex sets are defined corresponding to the global and local constraints:

$$\Pi_1 = \left\{ P \in \mathbb{R}^M : \sum_{i=1}^{M} \mu_i P_i = \xi \right\}, \Pi_2 = \left\{ X \in \mathbb{R}^M : X_i \in \Omega_i \right\}$$

together with their indicator functions [24]

$$I_1(P) = \begin{cases} 0 : P \in \Pi_1 \\ \infty : P \notin \Pi_1 \end{cases} \quad I_2(X) = \begin{cases} 0 : X \in \Pi_2 \\ \infty : X \notin \Pi_2 \end{cases}$$

Hence, (25) can be rewritten as follows:

$$\min \sum_{i=1}^{M} f_i(P_i) + I_1(P) + I_2(X) \tag{28}$$

$$\text{s.t.} P - X = 0 \tag{29}$$

Because the indicator function of a closed, non-empty convex set is proper, closed, and convex [25], the cost functions in (28) are also proper, closed, and convex with respect to P and X. Hence, (28) and (29) is exactly in a 2-block ADMM form [24]. Next, an augmented Lagrangian associated with (28) is defined as follows:

$$L_\rho(P, X, u) \triangleq \sum_{i=1}^{M} f_i(P_i) + I_1(P) + I_2(X) + \frac{\rho}{2}\|P - X + u\|_2^2$$

where $\rho > 0$ is a scalar penalty parameter and u is called the scaled dual variable or scaled Lagrange multiplier [24]. Subsequently, the optimization problem (28) and (29) can be iteratively solved where at each iteration $k = 1, 2, \ldots,$ the variables P, X, u are updated by [24]:

$$P^{k+1} = \underset{P}{\operatorname{argmin}}\, L_\rho\left(P, X^k, u^k\right) \tag{30}$$

$$X^{k+1} = \underset{X}{\operatorname{argmin}}\, L_\rho\left(P^{k+1}, X, u^k\right) \tag{31}$$

$$u^{k+1} = u^k + P^{k+1} - X^{k+1} \tag{32}$$

This algorithm is stopped if the criteria $\|r^k\|_2 \le \varepsilon^{\text{pri}}$ and $\|s^k\|_2 \le \varepsilon^{\text{dual}}$ are both satisfied, where $r^k \triangleq P^k - X^k$ and $s^k \triangleq X^k - X^{k-1}$ are the primal and dual residuals at iteration k; $\varepsilon^{\text{pri}} > 0$ and $\varepsilon^{\text{dual}} > 0$ are called primal and dual feasibility tolerances that can be chosen by

$$\varepsilon^{\text{pri}} = \sqrt{M}\varepsilon^{\text{abs}} + \varepsilon^{\text{rel}}\max\left\{\|P^k\|_2, \|X^k\|_2\right\}, \varepsilon^{\text{dual}} = \sqrt{M}\varepsilon^{\text{abs}} + \varepsilon^{\text{rel}}\|\rho u^k\|_2$$

with $\varepsilon^{\text{abs}} > 0$ and $\varepsilon^{\text{rel}} > 0$ which are some absolute and relative tolerances suggested to be 10^{-3} or 10^{-4} [24]. In [26], those tolerances and stopping criteria are shown to be computed and verified in distributed manners.

In the following, the variables P, X, u will be updated in a distributed manner using MAS and consensus theory. It can be observed that u is updated in a decentralized fashion when P and X are already updated. Therefore, only the updates of P and X are introduced below.

3.3 *P*-update step

The update of P in (30) is in fact equivalent to solving the following convex optimization problem:

$$\min_P \sum_{i=1}^{M} f_i(P_i) + \frac{\rho}{2}\|P - X^k + u^k\|_2^2 \tag{33}$$

$$\text{s.t.} \sum_{i=1}^{M} \mu_i P_i = \xi \tag{34}$$

of which the strong duality holds [25]. Let λ^{k+1} be the Lagrange multiplier associated with (34) and $\overline{\lambda}^{k+1}$ its optimal value, at the iteration $k+1$ of the SDC-ADMM algorithm. Then the Karush-Kuhn-Tucker (KKT) conditions can be used to find P^{k+1} from the following equations:

$$\left.\frac{\partial f_i(P_i)}{\partial P_i}\right|_{P_i=P_i^{k+1}} + \rho\left(P_i^{k+1} - X_i^k + u_i^k\right) = \overline{\lambda}^{k+1}\mu_i \tag{35}$$

$$\sum_{i=1}^{M}\mu_i P_i^{k+1} = \xi \tag{36}$$

Since $f_i(P_i)$ are known, P_i^{k+1} can be easily determined from (35). To find $\overline{\lambda}^{k+1}$, the values of P_i^{k+1} computed from (35) are substituted into (36). Usually, $f_i(P_i)$ are chosen to be linear or quadratic functions for simplicity. Hence, the first term on the left-hand side of (35) is in the form of a linear function of P_i^{k+1}, denoted by $\widetilde{a}_i P_i^{k+1} + \widetilde{b}_i$. This leads to

$$\overline{\lambda}^{k+1} = \frac{\xi + \sum_{i=1}^{M}\hat{b}_i}{\sum_{i=1}^{M}\hat{a}_i} \tag{37}$$

where $\hat{a}_i \triangleq \frac{\mu_i^2}{\tilde{a}_i+\rho}$, $\hat{b}_i \triangleq \frac{\mu_i\rho\left(-X_i^k+u_i^k\right)+\mu_i\tilde{b}_i}{\tilde{a}_i+\rho}$. The optimal value of λ^{k+1} shown in (37) is a global variable because the information of all variables are needed. Therefore, in order to calculate it in a distributed manner, consensus theory is employed where each unit in the grid corresponding to a variable is represented by an agent which exchange its parameters $\hat{a}_i$ and $\hat{b}_i$ and ξ (if $\xi \neq 0$) with a few neighbors in each iteration k of the SDC-ADMM algorithm. This is introduced in the following theorem:

Theorem 1. Given a connected undirected communication graph $\mathcal{G}$ among agents, the initial conditions of agents are set to be

$$\begin{aligned} z_i(0) &= \left[\hat{b}_i, \hat{a}_i\right]^T, i = 1, \ldots, M \\ z_{M+1}(0) &= [\xi, 0]^T \end{aligned} \tag{38}$$

Consequently, the following consensus law is utilized:

$$z(l+1) = w_{ii}z_i(l) + \sum_{j\in\aleph_i} w_{ij}z_j(l) \tag{39}$$

where w_{ij} are Metropolis weights [27] defined by

$$w_{ij} = \begin{cases} 1/\left(1+\max\left(|\aleph_i|, |\aleph_j|\right)\right) & : j\in\aleph_i \\ 1-\sum\limits_{j\in\aleph_i} w_{ij} & : j=i \\ 0 & : \text{otherwise} \end{cases} \tag{40}$$

Then the consensus of all agents is achieved, where the consensus states are

$$z_{*,1} = \frac{\xi + \sum_{i=1}^{M}\hat{b}_i}{M+1}, z_{*,2} = \frac{\sum_{i=1}^{M}\hat{a}_i}{M+1}$$

Thus, the optimal Lagrange multiplier can be obtained in a distributed manner by

$$\overline{\lambda}^{k+1} = \frac{z_{*,1}}{z_{*,2}} \tag{41}$$

Proof: Similar to the proof of Theorem 1 in [26], so it is omitted here for brevity. ■

Remark 1: If $\xi(t) = 0$, then z_{M+1} is not needed, and hence, there are only M agents in the consensus problem in Theorem 1. Accordingly,

$$z_{*,1} = \frac{\sum_{i=1}^{M} \hat{b}_i}{M}, z_{*,2} = \frac{\sum_{i=1}^{M} \hat{a}_i}{M}$$

3.4 *X*-update step

The update of X in (31) is obtained by solving a constrained quadratic programming:

$$\min_X \left\| P^{k+1} - X + u^k \right\|_2^2 \tag{42}$$

$$\text{s.t.} X \in \Pi_2 \tag{43}$$

Further, (42) and (43) is decentralized because both the cost function (42) and the constraint (43) are individually decomposable. Therefore, each agent (unit) just needs to solve a local scalar-constrained quadratic programming:

$$\min_{X_i} \left(P_i^{k+1} - X_i + u_i^k \right)^2 \tag{44}$$

$$\text{s.t.} X_i \in \Omega_i \tag{45}$$

which is easily solved by existing methods.

For the DSWM problem, (45) is simply an interval, and X_i has an analytical solution [26]. Similar results are obtained for the PSwEV problem.

3.5 Algorithm summarization

The convergence of the proposed SDC-ADMM algorithm can be proved in a similar manner to that in [26]; hence it is ignored here for brevity. In the following, a summary for the algorithm is provided.

Algorithm 2: SDC-ADMM algorithm for solving (1).

While $1 \leq t \leq N$ **do**
  Each agent determines its local constraint set $\Omega_i(t)$;
  For $k = 1, \ldots,$ max_iter **do**
    %P-update step
    Each agent calculates $\hat{a}_i$ and $\hat{b}_i$, or ξ;
    All agents run the consensus law (39) with initial values (38), then compute $\overline{\lambda}^{k+1}$ by (41);
    Each agent updates its variable P_i^{k+1} as in (37);
    %X-update step
    Each agent updates its variable X_i^{k+1} by solving (44) and (45);
    %u-update step
    Each agent updates its variable u_i^{k+1} as in (32);
    %Check the stopping criteria
    if the stopping criteria are true **then**
      break;
    end
  end
end

3.6 Pricing mechanism

For several grid optimization problems represented in the form of (1), e.g., economic dispatch or power scheduling, DSWM, PSwEV, etc., an electricity pricing mechanism is needed to drive the electricity trading in the grid. Interestingly, the optimal Lagrange multiplier associated with the equality constraint (2) is often regarded as the market-clearing price. That will be proved under mild assumptions in the following:

Theorem 2. For a sufficiently small but positive ρ, i.e., $\rho \to 0$, the optimal energy price in the considering smart grid converges to λ^*, the optimal Lagrange multiplier corresponding to the constraint (34) or equivalently (26), as $k \to \infty$.

Proof: For the OEM problems in smart grids, the convex functions $f_i(x_i(t))$ in (1) and (25) represent the costs for power generations or for the consumers' satisfaction. Therefore, we can denote the following functions as the reward functions for agents:

$$W_i(x_i(t)) = p(t)\mu_i x_i(t) - f_i(x_i(t)) \tag{46}$$

where $p(t)$ is the energy price at time slot t. In the ADMM reformulation (28) and (29), $x_i(t)$ is replaced by $P_i^{k+1}(t)$. For simplicity, the time index is also dropped hereafter.

The optimal energy price in the system is achieved when the market is cleared, at which the marginal costs of all agents are the same. Hence,

$$\frac{\partial W_i\left(P_i^{k+1}\right)}{\partial P_i^{k+1}} = 0 \; \forall i = 1, \dots, M \tag{47}$$

In the proposed SDC-ADMM algorithm, this means

$$\frac{\partial f_i\left(P_i^{k+1}\right)}{\partial P_i^{k+1}} = p\mu_i \; \forall i = 1, \dots, M \tag{48}$$

Substituting (48) into (35), we have

$$p\mu_i + \rho\left(P_i^{k+1} - X_i^k + u_i^k\right) = \bar{\lambda}^{k+1}\mu_i \; \forall i = 1, \dots, M \tag{49}$$

If $\rho \to 0$, we obtain from (49) that

$$p = \bar{\lambda}^{k+1} \tag{50}$$

Thus, as long as the proposed SDC-ADMM algorithm converges, the optimal energy price p^* is equal to the optimal Lagrange multiplier $\lambda^* \triangleq \lim_{k\to\infty} \bar{\lambda}^{k+1}$. ■

4. Simulation results for test systems

In this section, simulation results for two problems, DSWM and PSwEV, are provided to illustrate the effectiveness of the proposed SDC-ADMM algorithm.

4.1 Test case 1: DSWM problem

For this problem, a modified IEEE 39-bus system (see **Figure 1**) is used as the test system where it is assumed that there are PV generation units in the system

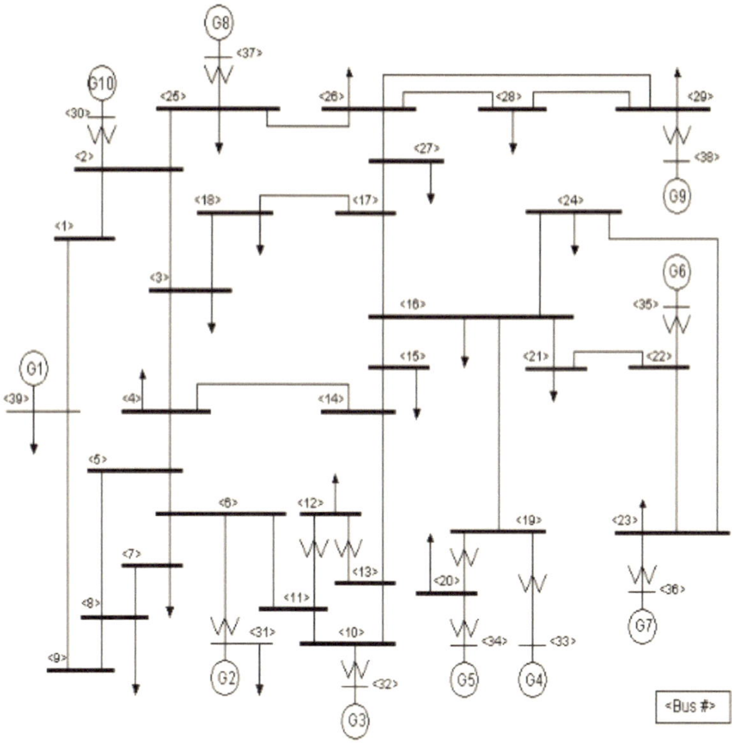

Figure 1.
IEEE 39-bus system.

with the total maximum output of 210 MW. The parameters of generators and demand units are taken from [26]. The average PV output curve, which is shown in **Figure 2**, is suitably scaled from the real PV data collected in New England [28]. Moreover, an average load profile taken from New England Independent System Operator [29] is utilized and properly scaled to obtain the time-varying upper bounds of demand units. This load profile has two demand peaks at 11:00 am and 8:00 pm, and the highest demand is at 8:00 pm, as seen in **Figure 3**.

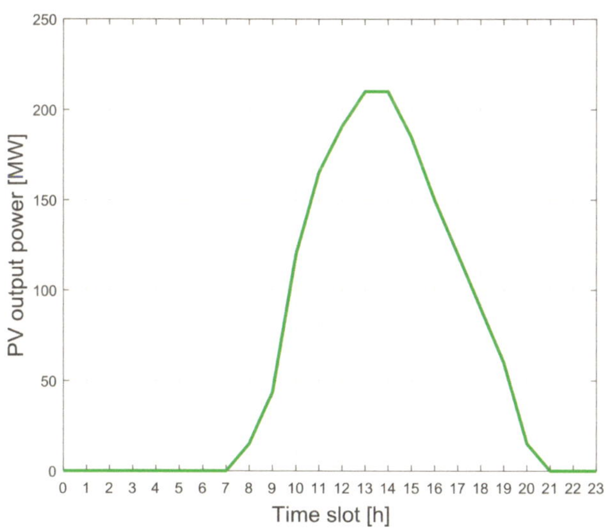

Figure 2.
PV output.

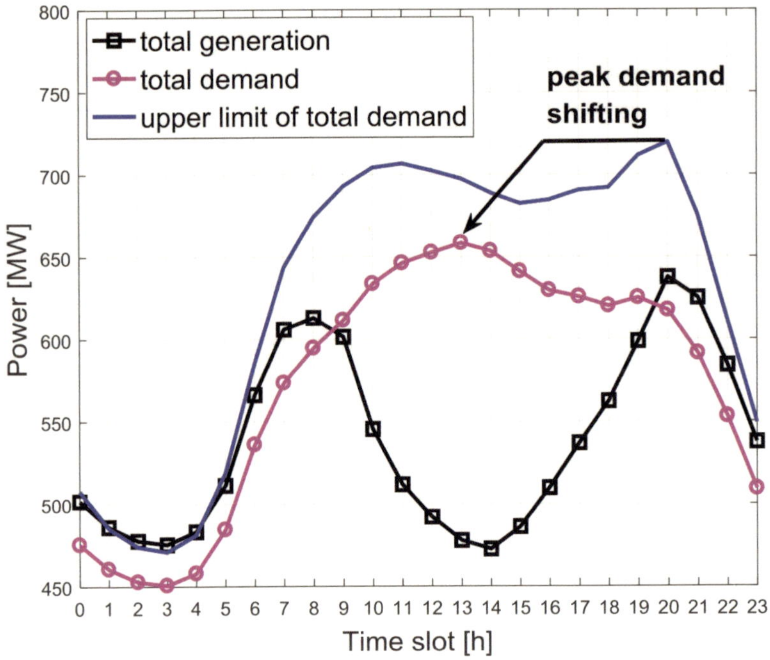

Figure 3.
Total CG power and demand.

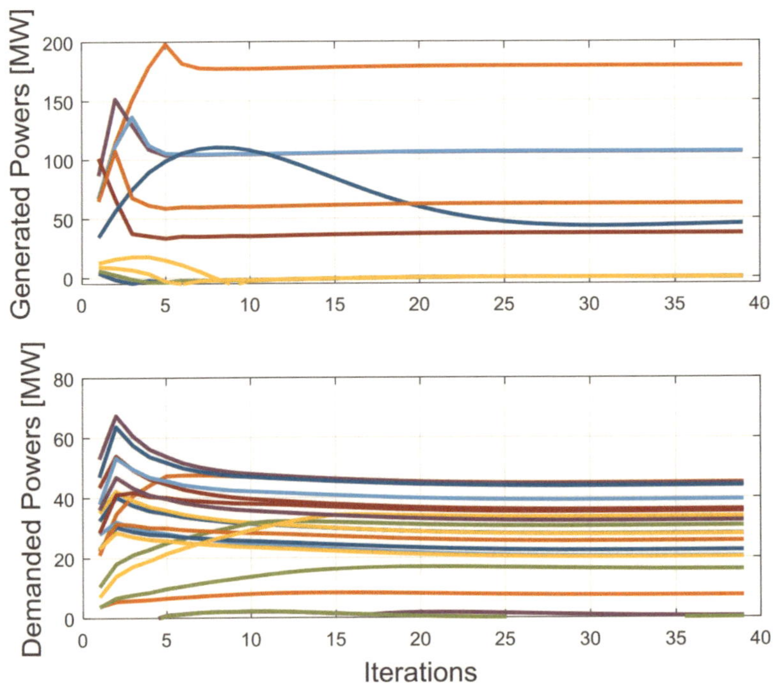

Figure 4.
Individual power profile.

Other parameters in the simulation are as follows. The absolute and relative tolerances are set to be $\varepsilon^{abs} = 10^{-4}$, $\varepsilon^{rel} = 10^{-3}$, while $\rho = 0.06$. Moreover, the power loss coefficients of generators and consumers are randomly generated within 10%. Then the simulation results are shown in **Figures 3–7**.

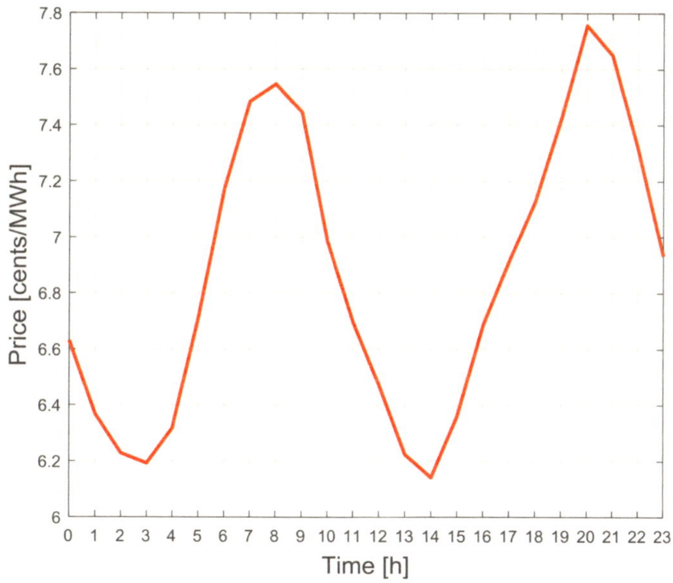

Figure 5.
Electricity price.

Figures 3 and **4** display the total and individual power generated and consumed in the test system which are obtained by the proposed SDC-ADMM algorithm. Thanks to PV energy, the peak demand is shifted from 8 pm to 1 pm at which the PV output is maximum, as observed in **Figure 3**.

Then the electricity price is exhibited in **Figure 5** showing that it is highest at 8 pm and lowest at 1–2 pm. This explains why the peak demand can be shifted. Next, the total welfare in the grid is shown in **Figure 6** where the maximum welfare is attained at 1 pm when the consumers use most energy and the CG units produce

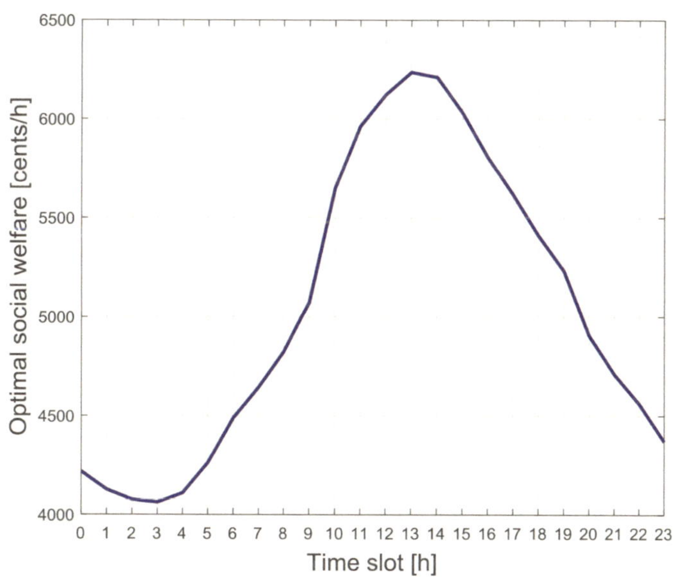

Figure 6.
Total social welfare.

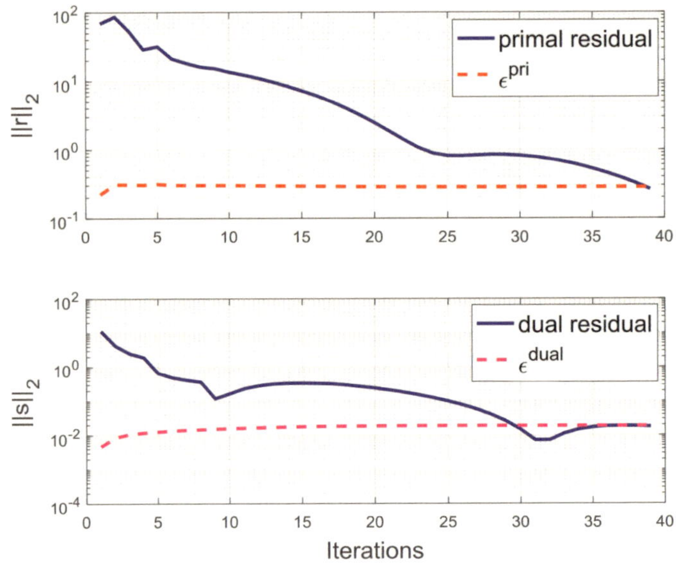

Figure 7.
Algorithm convergence.

least energy. Finally, **Figure 7** shows the convergence of the proposed SDC-ADMM algorithm.

4.2 Test case 2: PSwEV problem

In this test case, a microgrid having 2 DGs, 16 EVs, 4 load demands, 4 PV generations, and a microgrid operator is considered with a 3-day duration. The total number of agents is 27, in which the controllable agents are those for 2 DGs, 16 EVs, and the microgrid operator. The total PV curve and the total demand curve are given in **Figure 8**. The prescribed electricity transaction price $q(t)$ between the microgrid and the utility company is depicted in **Figure 9**.

The departure and arrival time and required traveling energies of 16 EVs are randomly generated from mix Gaussian distributions of this parameter of an actual

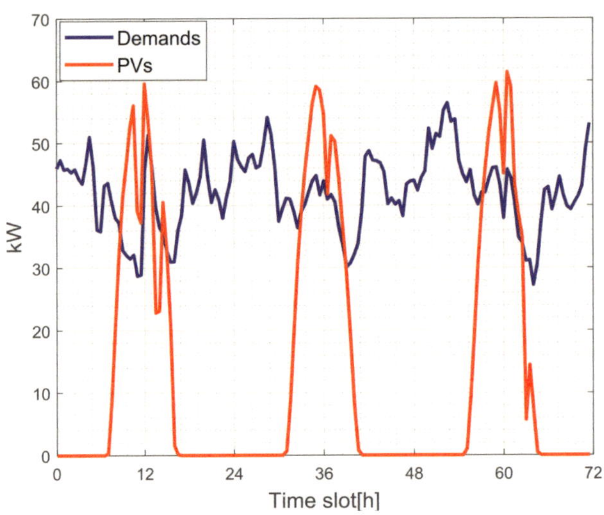

Figure 8.
PV and demand curves.

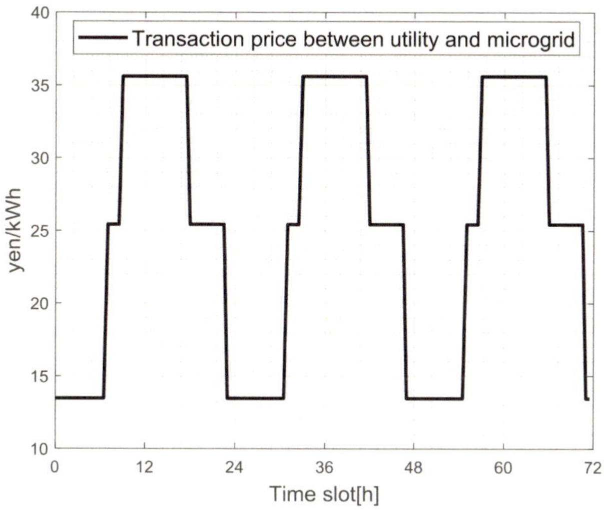

Figure 9.
Transaction price by utility.

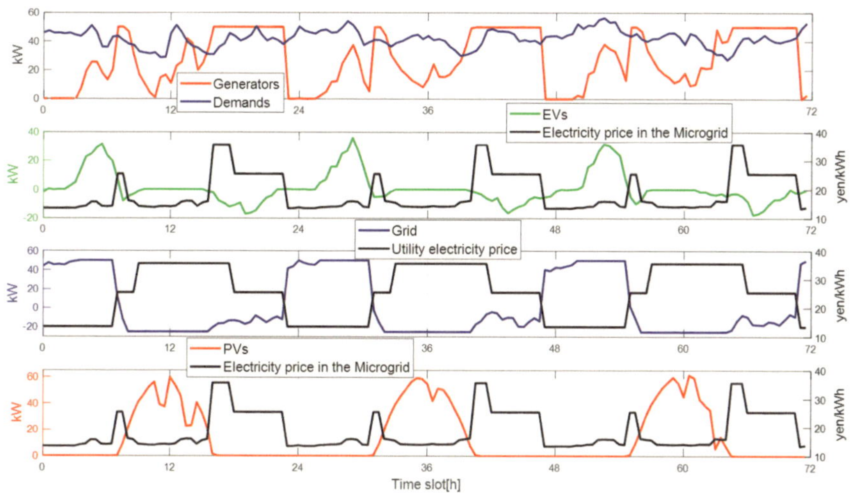

Figure 10.
DG power, EV charging power, and transaction price inside microgrid.

set of 1400 EVs. The maximum energy capacity of each EV battery is 17.6 kWh, and the SOC limits of each EV battery are set at 20 and 80% of the maximum energy capacity, respectively.

The absolute and relative tolerances are set to be $\varepsilon^{abs} = 10^{-4}$ and $\varepsilon^{rel} = 10^{-3}$. Next, $\rho = 0.06$. These parameters are the same with those in Section 4.1.

Consequently, the simulations for hundreds of different scenarios corresponding to the above random times are run with the proposed SDC-ADMM algorithm. The results for one specific scenario are then shown in **Figures 10–12**.

Figure 10 shows the benefit of utilizing PV generation and EV charging in the microgrid. First, DG output power is reduced around the time period of high PV output, while the load demand is quite high. Second, when PV output is high, the microgrid operator can sell the redundant electricity to the utility at the highest price. Last, even though EVs do not have much correlation to PV output due to the

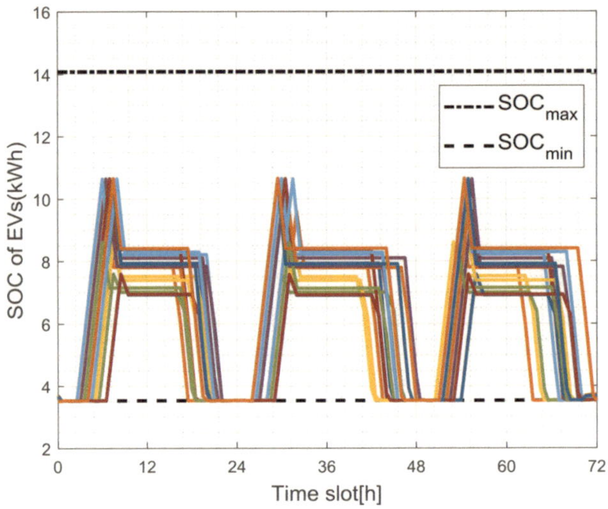

Figure 11.
The state of charge of all EVs.

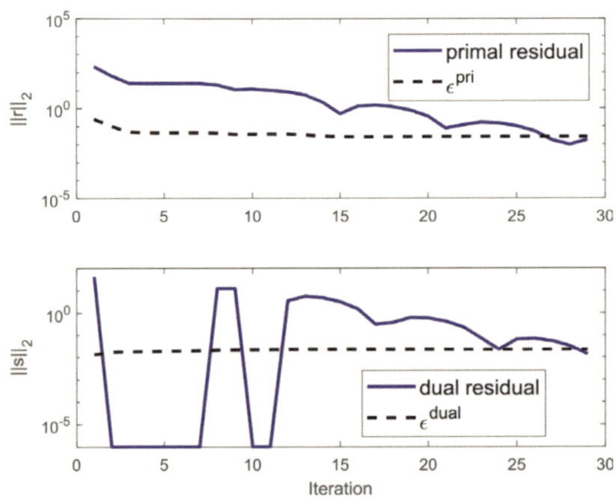

Figure 12.
SDC-ADMM convergence.

EV traveling times, they still benefit the microgrid with their optimal charging and discharging schedules in which the charging is executed at low-price time periods of utility, and vice versa, the discharging is made at high-price time periods of utility. Moreover, the discharging also provides microgrid electricity for selling to the utility at a high price.

Subsequently, the SOC profiles of 16 EVs are depicted in **Figure 11**. It can be observed that the EV batteries will be charged in the early morning when the electricity price is low, but not to the maximum allowed SOC, i.e., 80% of maximum battery capacity, because of the using of EV charging/discharging strategy in Algorithm 1 and the realistic EV data that only around 6 kWh is enough for each EV round-trip. Further, the charged/discharged SOCs of EVs are different due to their differences on charging/discharging times and required energy for traveling.

Finally, the convergence of the proposed SDC-ADMM algorithm in the PSwEV problem is shown in **Figure 11**, and the consensus processes in the distributed algorithm for calculating the optimal electric price inside the microgrid are displayed in **Figures 12** and **13**.

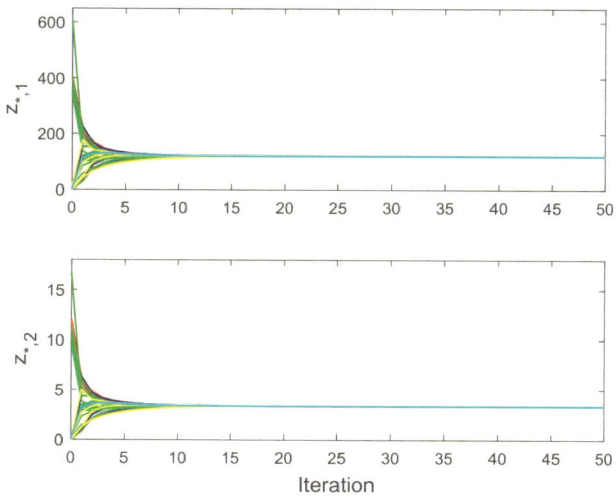

Figure 13.
Consensus for computing $\overline{\lambda}^{k+1}$.

5. Conclusions

In this chapter, a distributed optimization algorithm called sequential distributed consensus-based alternating direction method of multipliers (SDC-ADMM) is proposed for optimal energy management in smart grids. This algorithm is applicable to a broad class of linear or nonlinear constrained convex programming, of which two specific problems in smart grids have been studied in this chapter. The first problem DSWM tries to maximize the total social welfare in transmission grids in the presence of renewable energy and power losses, while the second problem PSwEV considers the power scheduling in distribution microgrids with renewable energy, electric vehicle as mobile storage, and a microgrid operator. It is then shown that the proposed SDC-ADMM algorithm works well for both problems, in which an optimal real-time electricity pricing scheme is derived as a part of the algorithm which facilitates demand response. Additionally, the existence of renewable energy and electric vehicles with suitable charging and discharging strategies is benefit to the grid for reducing the electricity price and the output power from nonrenewable energy generation.

Conflict of interest

The authors declare no conflict of interest.

Author details

Dinh Hoa Nguyen[1*], Huynh Ngoc Tran[2], Tatsuo Narikiyo[2]
and Michihiro Kawanishi[2]

1 International Institute for Carbon-Neutral Energy Research (WPI-I2CNER) and Institute of Mathematics for Industry (IMI), Kyushu University, Fukuoka, Japan

2 Department of Advanced Science and Technology, Toyota Technological Institute, Nagoya, Japan

*Address all correspondence to: hoa.nd@i2cner.kyushu-u.ac.jp

References

[1] U.S. Department of Energy. The Smart Grid: An Introduction [Internet]. Available from: https://www.energy.gov/oe/downloads/smart-grid-introduction-0 [Accessed: October 28, 2018]

[2] Erol-Kantarci M, Mouftah HT. Energy-efficient information and communication infrastructures in the smart grid: A survey on interactions and open issues. IEEE Communication Surveys and Tutorials. 2015;**17**(1): 179-197

[3] Vardakas JS, Zorba N, Verikoukis CV. A survey on demand response programs in smart grids: Pricing methods and optimization algorithms. IEEE Communication Surveys and Tutorials. 2015;**17**(1): 152-178

[4] Palensky P, Dietrich D. Demand side management: Demand response, intelligent energy systems, and smart loads. IEEE Transactions on Industrial Informatics. 2011;**7**(3):381-388

[5] Esther BP, Kumar KS. A survey on residential demand side management architecture, approaches, optimization models and methods. Renewable and Sustainable Energy Reviews. 2016;**59**: 342-351

[6] U.S. Department of Energy. Benefits of Demand Response in Electricity Markets and Recommendations for Achieving Them [Internet]. 2006. Available from: https://emp.lbl.gov/sites/all/files/report-lbnl-1252d.pdf [Accessed: October 28, 2018]

[7] Loia V, Vaccaro A. Decentralized economic dispatch in smart grids by self-organizing dynamic agents. IEEE Transactions on Systems, Man, and Cybernetics: Systems. 2014;**44**(4): 397-408

[8] Samadi P, Mohsenian-Rad AH, Schober R, Wong VWS, Jatskevich J. Optimal real-time pricing algorithm based on utility maximization for smart grid. In: Proceedings of the First IEEE International Conference on Smart Grid Communications; 4-6 October 2010; New York, USA: IEEE; 2010. pp. 415-420

[9] Samadi P, Mohsenian-Rad H, Schober R, Wong VWS. Advanced demand side management for the future smart grid using mechanism design. IEEE Transactions on Smart Grid. 2012; **3**(3):1170-1180

[10] Zhang W, Xu Y, Liu W, Zang C, Yu H. Distributed online optimal energy management for smart grids. IEEE Transactions on Industrial Informatics. 2015;**11**(3):717-727

[11] Rahbari-Asr N, Ojha U, Zhang Z, Chow MY. Incremental welfare consensus algorithm for cooperative distributed generation/demand response in smart grid. IEEE Transactions on Smart Grid. 2014;**5**(6): 2836-2845

[12] Li N, Chen L, Low SH. Optimal demand response based on utility maximization in power networks. In: Proceedings of IEEE Power Energy Society General Meeting (2011 IEEE PESGM); 24-29 July 2011; New York, USA: IEEE; 2011. pp. 1-8

[13] Xu Y, Yang Z, Gu W, Li M, Deng Z. Robust real-time distributed optimal control based energy management in a smart grid. IEEE Transactions on Smart Grid. 2017;**8**(4):1568-1579

[14] Zhao C, He J, Cheng P, Chen J. Consensus-based energy management in smart grid with transmission losses and directed communication. IEEE Transactions on Smart Grid. 2017;**8**(5): 2049-2061

[15] Jiang B, Farid AM, Youcef-Toumi K. Demand side management in a day-ahead wholesale market: A comparison of industrial & social welfare approaches. Applied Energy. 2015;**156**: 642-654

[16] Zhong W, Yu R, Xie S, Zhang Y, Yau DKY. On stability and robustness of demand response in V2G mobile energy networks. IEEE Transactions on Smart Grid. 2018;**9**(4):3203-3212

[17] Kamboj S, Kempton W, Decker KS. Deploying power grid-integrated electric vehicles as a multi-agent system. In: Proceedings of International Conference on Autonomous Agents and Multiagent Systems; 2 May 2011; Taiwan. Taipei; 2011. pp. 13-20

[18] Japanese Ministry of Economy, Trade and Industry. Automobile Industry Strategy. 2014. Available from: http://www.meti.go.jp/english/press/2014/1117_01.html [Accessed: October 28, 2018]

[19] Logenthiran T, Srinivasan D, Shun TZ. Demand side management in smart grid using heuristic optimization. IEEE Transactions on Smart Grid. 2012;**3**(3): 1244-1252

[20] Gudi N, Wang L, Devabhaktuni V. A demand side management based simulation platform incorporating heuristic optimization for management of household appliances. International Journal of Electric Power and Energy Systems. 2012;**43**(1):185-193

[21] Kempton W, Tomia J. Vehicle-to-grid power fundamentals: Calculating capacity and net revenue. Journal of Power Sources. 2005; **144**(268):268-279

[22] Vandael S, Holvoet T, Deconinck G, Kamboj S, Kempton W. A comparison of two GIV mechanisms for providing ancillary services at the University of Delaware. In: Proceedings of IEEE International Conference on Smart Grid Communications; 21 October 2013; Canada. Vancouver: IEEE; 2013. pp. 211-216

[23] Kumar Debnath U, Ahmad I, Habibi D, Yousuf Saber A. Improving battery lifetime of gridable vehicles and system reliability in the smart grid. IEEE Systems Journal. 2015;**9**(3):989-999

[24] Boyd S, Parikh N, Chu E, Peleato B, Eckstein J. Distributed optimization and statistical learning via the alternating direction method of multipliers. Foundation and Trends in Machine Learning. 2011;**3**(1):1-122

[25] Boyd SP, Vandenberghe L. Convex Optimization. Cambridge, UK: Cambridge University Press; 2004. 716p. DOI: 10.1017/CBO9780511804441

[26] Nguyen DH, Narikiyo T, Kawanishi M. Optimal Demand Response and Real-time Pricing by a Sequential Distributed Consensus-based ADMM Approach. IEEE Transactions on Smart Grid. 2018;**9**(5):4964-4974

[27] Xiao L, Boyd S, Lall S. A scheme for robust distributed sensor fusion based on average consensus. In: Proceedings of the Fourth International Symposium on Information Processing in Sensor Networks (IPSN 2005); 15 April 2005; New York, USA: IEEE; 2005. pp. 63-70

[28] Bzura JJ. Residential photovoltaics: The New England experience builds confidence in PV. Photovoltaics new opportunities utilities. Technical Report E91002168 [Internet]. Washington, DC, USA: Office Sci. Tech. Inf., U.S. Department of Energy; 1991. Available from: http://www.osti.gov/scitech/servlets/purl/5253894 [Accessed: October 28, 2018]

[29] New England ISO. Energy, Load, and Demand Reports [Internet]. 2009. Available from: https://www.iso-ne.com/isoexpress/web/reports/pricing/-/tree/zone-info [Accessed: October 28, 2018]

Chapter 4

Voltage Regulation in Smart Grids

Maher Azzouz

Abstract

The intermittent nature of renewable power sources (RES) can significantly change the voltage profile of smart grids and adversely impact the conventional voltage control devices such as tap-changing transformers and capacitor banks. Furthermore, the growing penetration of plug-in electric vehicles (PEVs) can add high stress on voltage control devices due to the PEV stochastic and concentrated power profiles. Such power profiles may lead to high maintenance costs and reduced lifetimes for voltage control devices and limit actions on accommodation of high penetration levels of RES and PEVs. This chapter explains the basic background of voltage regulation in smart grids. The typical approaches, which are employed by utilities for voltage regulation, are reviewed. Then, the impact of RES and PEVs on voltage regulation is analyzed. Lastly, remedies for voltage violations in smart grids, such as optimal reactive power control and coordination between voltage control devices, are discussed.

Keywords: distributed generation, active distribution networks, voltage regulation, electric vehicles, onload tap changers

1. Introduction

Ongoing rapid advances in power electronics and communication technologies are facilitating the development of smart grids that are characterized by hosting distributed generation units (DGs). There are many benefits of smart grids such as enhancing power reliability and power quality, improving safety and cyber security, maximizing energy utilization and efficiency, environment protection and conservation, and increasing financial revenues [1]. For instance, as reported in [2], smart grids could reduce greenhouse gas emissions by up to 18%. The integration of renewable-based DGs alters distribution systems so that rather than having passive structures, with unidirectional power flow, they become active distribution networks (ADNs), with multidirectional power flow. Voltage regulation is considered one of the main operational challenges that accompany high penetration levels of renewable-based DGs. RES, such as wind and solar energy, can significantly change the voltage profile of smart grids and interact negatively with conventional schemes of controlling onload tap changers (OLTCs). Another factor is the growing penetration of PEVs, which creates additional stress on voltage control devices due to their stochastic and concentrated power profiles. These combined generation and load power profiles can lead to overvoltages, undervoltages, high system losses, excessive tap operation, infeasible solutions (hunting) with respect to OLTCs, and/or limits on the integration of either PEVs or RES.

2. Background

One of the main objectives of electric utilities is to maintain the grid voltage within standard levels to guarantee customers' satisfaction. Many equipment are deployed in the grid, such as OLTCs and capacitor banks, to properly fulfill this objective. In general, voltage regulation in a smart grid can be classified as local or communication-assisted [3]. The local voltage regulation is referred to as the conventional control, where the reference values and measurements for voltage control are locally determined. In this section, the fundamentals of the conventional control of OLTC and inverter-interfaced DGs are explained.

2.1 OLTC conventional controller

OLTCs are one of the main voltage regulators in distribution systems. A tap changer equipped with an automatic control system usually regulates the transformer's secondary voltage to maintain an acceptable voltage near the load center. The tap changer mechanically varies the tap position from zero (no voltage compensation) to N_{max} (maximum voltage compensation). This allows for changing the transformer's turn ratio in discrete steps; each step adjustment may take 3–20 seconds. **Figure 1** displays the schematic diagram of an OLTC, where the tap changer is placed on the primary side. The tap changer is usually installed on the high voltage side of the transformer (i.e., the primary side in **Figure 1**) because of the low current of that side. A reversing switch is employed to change the polarity of the tap winding for positive and negative voltage compensation. In steady state, the transformer model can be given by

$$V_2 = \frac{V_1}{a} - I_2 Z_T(a) \tag{1}$$

where V_1 and V_2 are primary and secondary voltages, respectively, I_2 is secondary current, a denotes the transformer's tap ratio, and $Z_T\,(a)$ is the transformer

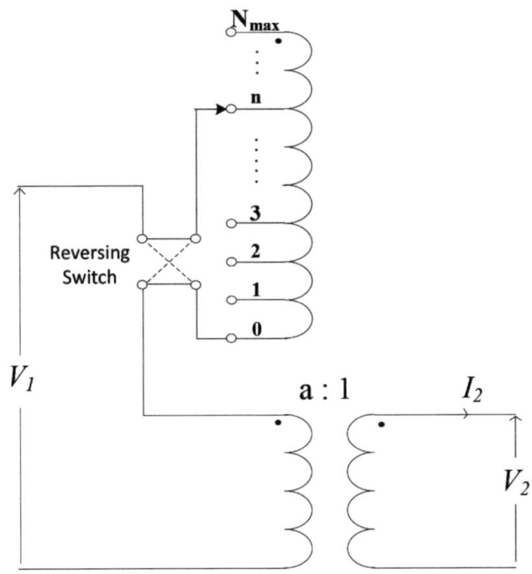

Figure 1.
Transformer with tap changer.

series impedance (referred to the secondary side). The tap changer allows a to vary linearly; thus it can be expressed in terms of the nominal turn ratio (i.e., $a_0 = 1.0$):

$$a = a_0 + n_i \Delta a \tag{2}$$

where Δa is the change in a as a result of a step change in the tap position n_i, which is given by

$$n_i = n_{i-1} + \Delta n \tag{3}$$

where n_i is the present tap position and $n_i - 1$ is the previous tap position. Δn is the change in the tap position which is defined by

$$\Delta n = \begin{cases} 0 & for \quad t \leq T_m \quad : \quad b = arbitrary \\ 1 & for \quad t > T_m \quad : \quad b = -1 \\ -1 & for \quad t > T_m \quad : \quad b = 1 \end{cases} \tag{4}$$

where T_m denotes mechanical time delay which is required by the motor driver unit to change the tap position by only one step and b is the control signal that is applied to the tap-changing mechanism (see **Figure 2**) and is given by

$$b = \begin{cases} 0 & for \quad t \leq T_d \quad : \quad e = arbitrary \\ 1 & for \quad t > T_d \quad : \quad e = 1 \\ -1 & for \quad t > T_d \quad : \quad e = -1 \end{cases} \tag{5}$$

where T_d is the time delay introduced by the OLTC controller and e is the output of the hysteresis controller, that is,

$$e = \begin{cases} 0 & for \quad |\Delta V| \leq DB \\ 1 & for \quad \Delta V > DB \\ -1 & for \quad \Delta V < -DB \end{cases} \tag{6}$$

The mechanical time delay (T_m) has a constant value; usually, it varies from 3 to 10 seconds, and ΔV is the voltage error. In some models of the OLTC, the controller time delay (T_d) is considered constant:

$$T_d = \tau_0 \tag{7}$$

However, T_d is usually variable and depends on the voltage error and the controller's dead band DB [4]:

$$T_d = \tau_0 \frac{DB}{|\Delta V|} \tag{8}$$

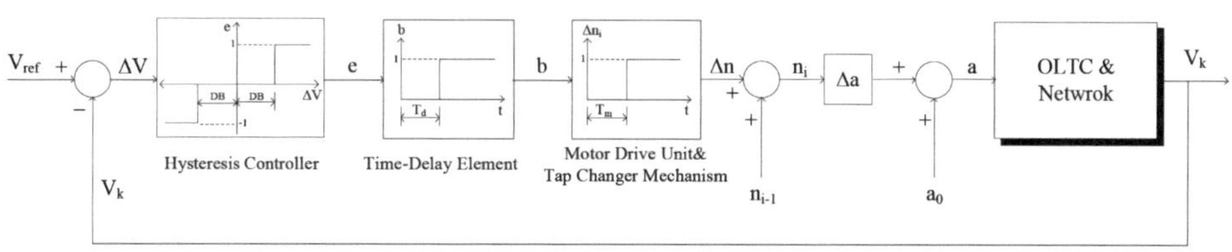

Figure 2.
The schematic diagram of the conventional discrete control of the OLTC.

T_d is inversely proportional to the voltage error ΔV to avoid unnecessary operation during transient voltages and temporary load disturbances. DB must be greater than Δa to ensure stability when the regulated voltage approaches the reference valueV_{ref}; otherwise, the OLTC will suffer from hunting. V_{ref} is typically between 0.95 p.u. and 1.0 p.u. since it determines the steady-state voltage near the load center (V_k).

2.2 Conventional voltage control of inverter-based DGs

Energy processing strategies for inverter-based DGs typically involve two cascaded loops: inner and outer. The inner loop is a current control loop, which regulates the DG inverter current in the $d-q$ reference frame. The outer control loop, on the other hand, can fulfill different control objectives depending on the hosing grid, such as voltage regulation and power management control.

Figure 3 shows an inverter-based DG that is controlled in the current injection mode, which is the typical strategy adopted with RES. A DG inverter model in the $d-q$ synchronous frame represents the dynamics of the interfacing LC filter [5], that is,

$$L_f \frac{dI_d}{dt} = -R_f I_d + V_d - V_{od} + \omega L_f I_q \tag{9}$$

$$L_f \frac{dI_q}{dt} = -R_f I_q + V_q - V_{oq} - \omega L_f I_d \tag{10}$$

$$C_f \frac{dV_{od}}{dt} = I_d - I_{od} + \omega C_f Voq \tag{11}$$

$$C_f \frac{dV_{oq}}{dt} = I_q - I_{oq} - \omega C_f Vod \tag{12}$$

$$\omega = \frac{d\theta}{dt} \tag{13}$$

where I_{dq} and I_{odq} represent the $d-q$components of the inverter output current and DG current at the PCC, respectively; V_{dq} and V_{odq} are the components of the

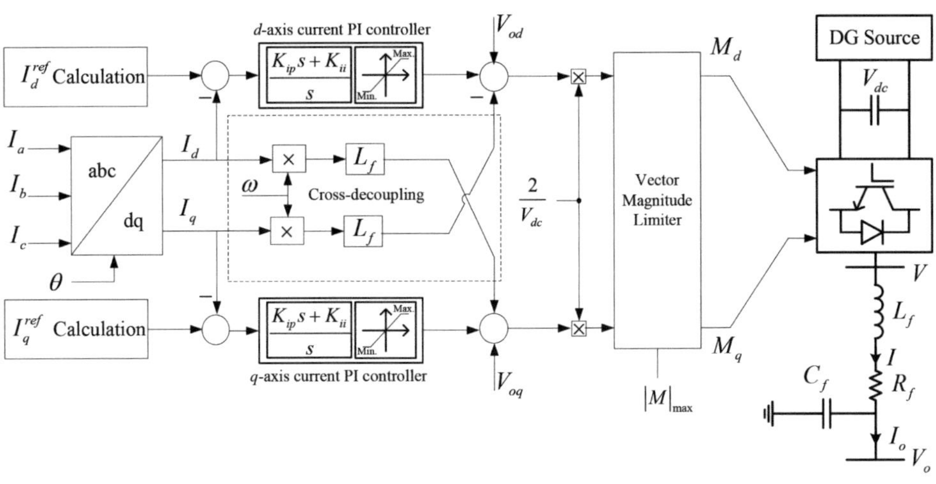

Figure 3.
Power circuit and current control diagram for inverter-based DG.

inverter terminal voltage and DG voltage at the PCC, respectively; R_f, L_f, and C_f are the resistance, inductance, and capacitance of the DG interfacing filter, respectively; and ω is the grid frequency.

The current equations, that is, (9) and (10), are coupled through the $\omega L_f I_q$ and $-\omega L_f I_d$ terms. For independent control of both the I_d and I_q currents, the decoupled terms must be eliminated, which can be accomplished if new variables V'_d and V'_q are defined by

$$V'_d = V_d - V_{od} + \omega L_f I_q \quad (14)$$

$$V'_q = V_q - V_{oq} - \omega L_f I_d \quad (15)$$

Substituting from (14) and (15) into (9) and (10) yields.

$$L_f \frac{dI_d}{dt} = -R_f I_d + V'_d \quad (16)$$

$$L_f \frac{dI_q}{dt} = -R_f I_q + V'_q \quad (17)$$

Equations (16) and (17) represent decoupled first-order differential equations for I_d and I_q, respectively. Therefore, PI current controllers can be designed based on the transfer functions derived from (16) and (17). If the gains of the PI current controllers are selected as

$$\begin{cases} K_{ip} = \dfrac{L_f}{\tau_i}, \\ K_{ii} = \dfrac{R_f}{\tau_i} \end{cases} \quad (18)$$

then the equivalent closed-loop transfer functions for the current loops can be given by

$$\frac{I_d(s)}{I_d^{ref}(s)} = \frac{I_q(s)}{I_q^{ref}(s)} = \frac{1}{\tau_i s + 1} \quad (19)$$

where τ_i is the time constant of the closed-loop system. The vector magnitude of the reference current $\left(\sqrt{\left(i_d^{ref}\right)^2 + \left(i_q^{ref}\right)^2}\right)$ should be limited according to the maximum allowable current, typically 20% greater than the rated current of the inverter [5], in order to provide overcurrent protection. The vector magnitude of the modulation index $\left(|M| = \sqrt{M_d^2 + M_q^2}\right)$ should also be limited to $|M|_{max} = 1.0$ p.u. so that DG operates in a linear modulation region. **Figure 3** shows the vector magnitude limiter, which implies a limit on the magnitude of $|M|$ without a change in the phase angle between M_d and M_q.

Determination of the reference currents, I_d^{ref} and I_q^{ref}, is usually affiliated to the DG output power. In the synchronous $d - q$ frame, the output active P_G and reactive Q_G powers are given by

$$P_G = 1.5(V_{od} I_d + V_{oq} I_q) \quad (20)$$

$$Q_G = 1.5(V_{oq} I_d - V_{od} I_q) \quad (21)$$

For decoupled control of the active and reactive powers, a phase-locked loop (PLL) should be used for aligning V_{od} with the output voltage of phase a so that $V_{oq} = 0$ [6]. To determine I_d^{ref}, a maximum power point tracking (MPPT) algorithm along with a DC link voltage controller is employed [7]. On the other hand, I_d^{ref} can be calculated, as in (22), to regulate the output reactive power, thus, providing voltage support.

$$I_d^{ref} = \frac{Q_G^{ref}}{1.5V_{od}} \tag{22}$$

where Q_G^{ref} is the reference value of the DG reactive power which can be determined using an outer voltage control loop or received from a communication-assisted voltage regulation scheme [8].

2.3 DG contribution to voltage violation

Usually, distribution networks have unidirectional power flow from the substation to customers. This leads to a descending voltage profile which may only suffer from an undervoltage near the load center, a problem typically tackled using OLTC and capacitor banks. On the other hand, DG integration into distribution networks makes the power flow bi-directional; thus an overvoltage problem may also occur. To understand the impact of DGs on the system voltage, the simplified distribution network in **Figure 4(a)** is used, where a DG is connected at a load bus. In this figure, R and X are the feeder resistance and reactance, respectively; I_R is the feeder current; V_1, V_2 are the primary and secondary voltage of the distribution transformer, respectively; and V_G is the DG output voltage. The phasor diagram of the simplified distribution network is shown in **Figure 4(b)**, in which δ is the power angle and φ is the phase shift between V_G and I_R.

Using the phasor diagram, the relation between the DG voltage and V_2 can be formulated:

$$\begin{aligned} V'_G &= [V_2 + I_R R\cos(\varphi) + I_R X \sin(\varphi)] \\ &= V_G \cos(\delta) \end{aligned} \tag{23}$$

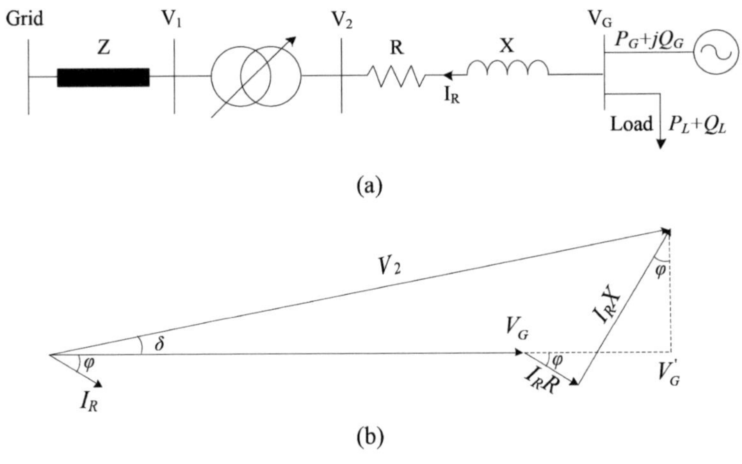

Figure 4.
Simplified distribution network with DG. (a) Single-line diagram and (b) phasor diagram.

The power angle (δ) is very small; hence (23) can be approximated by

$$V_G \approx V_2 + I_R R \cos(\varphi) + I_R X \sin(\varphi) \tag{24}$$

Therefore, the voltage rise caused by the DG, that is, $\Delta V_G = V_G - V_2$, is given by.

$$\Delta V_G \approx I_R R \cos(\varphi) + I_R X \sin(\varphi) = \frac{P_R R + Q_R X}{V_G} = \frac{(P_G - P_L)R + (Q_G - Q_L)X}{V_G} \tag{25}$$

where P_G and Q_G are the DG output active and reactive powers, respectively, while P_L and Q_L are the load active and reactive powers, respectively. It can be seen from (25) that the highest overvoltage happens when the DG generates its maximum power during a light load condition. This problem is mainly associated with the excessive reverse power flow caused by the DG.

3. Communication-assisted voltage regulation

Due to the intermittency of RES and PEVs, the conventional control schemes for OLTC and DGs fail to provide proper voltage regulation. This shortcoming can be compensated using communication-assisted voltage regulation schemes. In the literature, the communication-assisted schemes fall under two approaches: distributed and centralized [3]. Both approaches involve investment in communication links and remote terminal units. The distributed (intelligent) approach is considered to be an expert-based control or model-free approach, which coordinates a variety of voltage control devices with the goal of providing effective and nonoptimal voltage regulation with fewer communication requirements [9]. On the other hand, the centralized approach relies on a central point that monitors the system status and optimizes the operation of voltage control equipment. Typically, a centralized optimization problem is solved to dispatch the reactive power of different voltage control equipment based on (i) load forecasting and (ii) generation monitoring. Several solutions have been proposed in the literature to provide optimal reactive power dispatch for DGs [10–12]. In this section, the role of PEVs in optimal voltage regulation is explained as in [13].

3.1 PEV impact on voltage regulation

Figure 5 represents a simplified multi-feeder distribution network connected to a substation through an OLTC. The test network has a photovoltaic (PV)-based DG and a PEV parking lot, which are connected at different feeder terminals. Following the derivation of (25), the per-unit voltage deviation for both DG and PEV busses can be approximated by

$$\begin{aligned} \Delta V_{PV} &\approx (P_{PV} - P_{L_1})R_{f_1} + (Q_{PV} - Q_{L_1})X_{f_1} \\ \Delta V_{EV} &\approx -(P_{EV} + P_{L_2})R_{f_2} - (Q_{EV} + Q_{L_2})X_{f_2} \end{aligned} \tag{26}$$

where P_{PV}, P_{EV}, and P_L are DG, PEV, and load active powers, respectively, and Q_{PV}, Q_{EV}, and Q_L are DG, PEV, and load reactive powers, respectively.

Equation (26) shows that two worst-case scenarios may occur: (i) overvoltage, when the DG generates its maximum power during light loads and (ii) undervoltage, during a peak load demand and low DG output. The integration

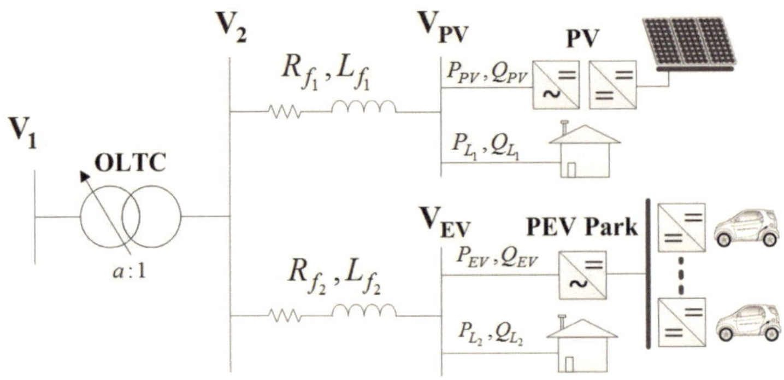

Figure 5.
Simplified distribution network with DG and PEVs.

of DGs changes the voltage profile significantly and complicates the voltage regulation. This is due to two reasons: (i) the voltage trend not descending from the substation to the feeder terminal, thereby invalidating the target point (reference) and (ii) the voltage estimation, based on local measurements, becoming inaccurate because of the stochastic power natures of RES and EVs [13]. Moreover, the stochastic power nature of EVs makes the voltage estimation inferior and aggravates the undervoltage problem. Therefore, OLTCs may suffer from wear and tear due to excessive operations. This problem worsens when feeders suffer from overvoltage due to high DG penetration, while others suffer from undervoltage during high demand, such as PEV charging. In this instance, the OLTC will have two contradicting solutions. Increasing the transformer's secondary voltage mitigates the undervoltage problem at the expense of the system's overvoltage and vice versa. **Figure 6** shows two daily power profiles for uncontrolled[1] PEV charging demand and a PV-based DG. The PEV demand is generated based on practical arrival/departure times from the Toronto Parking Authority (TPA), Toronto, Canada. Since the power profiles of commercial parking lots and PV-based DGs naturally coincide, there is a high chance that the system simultaneously suffers from overvoltage and undervoltage. A partial solution for this problem can be realized if a centralized-based controller for the OLTC exploits the system's maximum and minimum voltages. However, this controller may not prevent the OLTC hunting

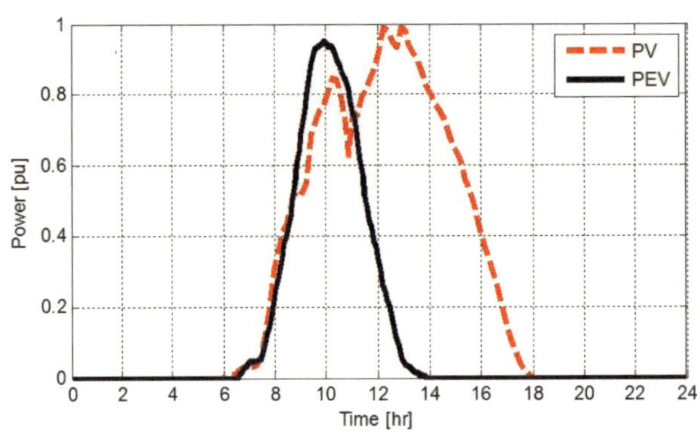

Figure 6.
DG and PEV power profiles.

[1] In uncontrolled charging schemes, PEVs start charging as soon as they are plugged in.

problem during high PV power generation and peak EV demand, resulting in excessive tap operation.

For that reason, the power electronic converters that interface DGs and PEVs should be utilized in voltage regulation. The DG can support the voltage regulation through two options: (i) absorbing reactive power and/or (ii) curtailment of active power. The first option is preferred since active power curtailment represents an energy waste. However, the capacity of the DG converter may limit the reactive power support and force the second option. To increase the reactive power support, the interfacing converter of the PEV can be employed to inject its surplus reactive power, thus reducing the DG active power curtailment [13]. A novel optimal coordinated voltage regulation scheme is presented to coordinate PEV, DG, and OLTC to achieve optimal voltage regulation and satisfy the self-objectives of each voltage control device.

3.2 OLTC centralized control

As shown in **Figure 7**, the OLTC is represented by a π-circuit model [14]. The taps are assumed to be at the primary side (high voltage). Subsequently, the OLTC secondary voltage and current can be calculated by

$$\begin{bmatrix} V_{(1,t)} \\ I_{(1,t)} \end{bmatrix} = \begin{bmatrix} \frac{1}{a} & -\frac{a}{\mathrm{Y}_T} \\ 0 & -a \end{bmatrix} \begin{bmatrix} V_{(0,t)} \\ I'_{(0,t)} \end{bmatrix} \tag{27}$$

where Y_T is the transformer series admittance, a is the turns ratio given in (2), and t denotes the time instant. To take the physical busses into account, (27) can be rewritten as

$$\begin{bmatrix} I_{(0,t)} \\ I_{(1,t)} \end{bmatrix} = \underbrace{\begin{bmatrix} g_F + jb_\mu + \frac{Y_T}{a^2} & -\frac{Y_T}{a} \\ -\frac{Y_T}{a} & Y_T \end{bmatrix}}_{\mathrm{Y_{OLTC}}} \begin{bmatrix} V_{(0,t)} \\ V_{(1,t)} \end{bmatrix} \tag{28}$$

where $\mathrm{Y_{OLTC}}$ is the OLTC Y-bus admittance matrix, which represents the OLTC admittance in the power flow equations.

The conventional OLTC controller, shown in **Figure 8(a)**, is modified to emulate an adaptive reference by considering the system's minimum and maximum voltages, that is, $V^{sys}_{\min}$ and $V^{sys}_{\max}$, respectively. This modification forms the centralized OLTC controllers (COC) proposed in [13] and shown in **Figure 8(b)**, which

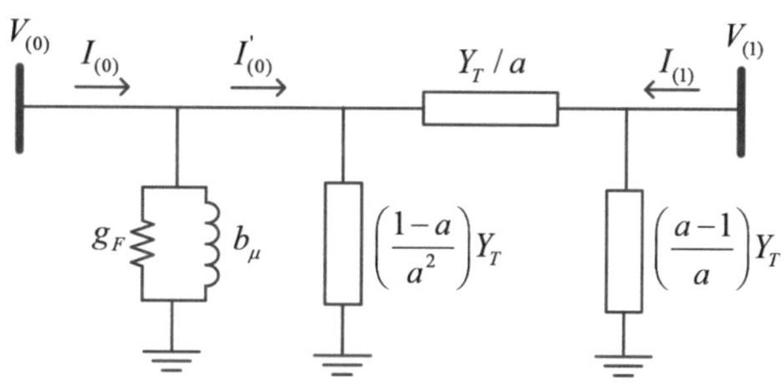

Figure 7.
Equivalent π-circuit model of OLTC.

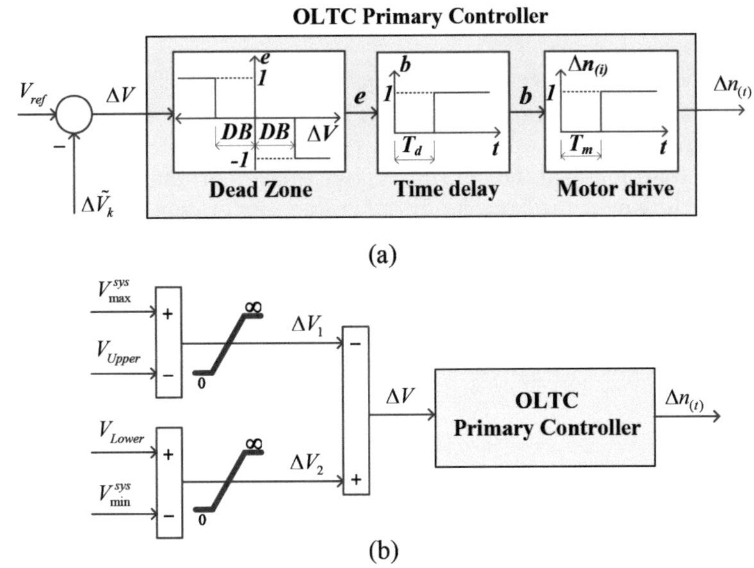

Figure 8.
The OLTC control: (a) conventional local controller and (b) centralized OLTC controller (COC).

allows the OLTC to deal with multiple feeders with voltage problems resulted from DGs and PEVs. The COC emulates an adaptive reference because the error ΔVchanges such that $V^{sys}_{\min}$and $V^{sys}_{\max}$ are within the standard limitsV_{Upper} and V_{Lower} (i.e., 1.05 and 0.95 p.u.), respectively. $V^{sys}_{\min}$ and $V^{sys}_{\max}$ can be estimated using the state estimation algorithm in [15] or attained through the central voltage control unit explained in the next subsection.

In the case of an overvoltage, ΔV is negative, and the primary controller decreases the transformer secondary voltage (by increasing the tap position) and vice versa. During normal conditions, ΔV is zero because both $V^{sys}_{\min}$ and $V^{sys}_{\max}$ are within the standard limits; thus, the tap position remains unchanged. When both overvoltage and undervoltage occur simultaneously, the COC should be disabled to avoid hunting [16]. In the next subsection, the roles of DGs and PEVs in voltage regulation are explained to mitigate the shortcoming of the COC.

3.3 Optimal coordinated voltage regulation

Both power electronic converters of PEVs and DGs can support the grid with reactive power to relax the OLTC. A vehicle-to-grid reactive power support (V2GQ) strategy is proposed in [13] to incorporate PEVs and DGs in voltage regulation, as shown in **Figure 9**. The main difference between vehicle-to-grid (V2G) strategies and the V2GQ is that the latter injects only reactive power to the grid. Thus, it preserves the battery life of PEV, that is, the highest priority of the vehicles' owners. Nevertheless, the V2GQ cannot be flexibly employed in power management because PEVs do not export active power to the grid. V2G strategies are avoided in voltage regulation to elongate the battery life, which is considered a priority in this study. The V2GQ comprises a three-stage nonlinear programming, in which Stage (I) aims at maximizing the energy delivered to PEVs, Stage (II) minimizes the DG active power curtailment, and Stage (III) minimizes the voltage deviations. The COC is coordinated such that it acts after Stage (III) to ensure that all bus voltages are within the standard limits.

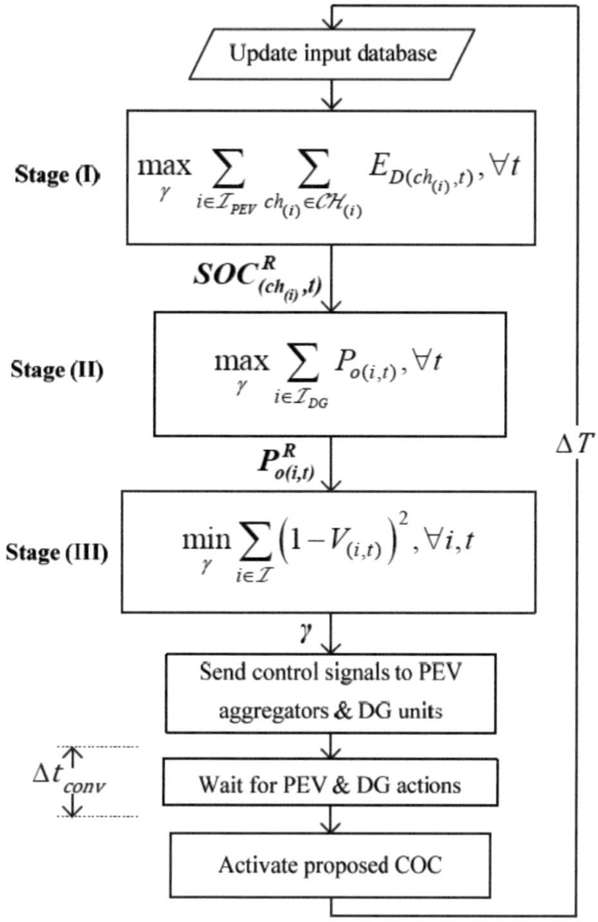

Figure 9.
PEV/DG voltage support scheme.

3.3.1 Problem formulation of stage (I)

The main objective of Stage (I) is maximizing the energy delivered to PEV owners, that is,

$$\max_{\gamma} \sum_{i \in \mathcal{I}_{PEV}} \sum_{ch_{(i)} \in \mathcal{CH}_{(i)}} E_{D\left(ch_{(i)},t\right)} \qquad \forall t \tag{29}$$

where $E_{D(\mathrm{ch(i)},t)}$ is the energy delivered to the PEV connected to charger $ch_{(\mathrm{i})} \in \mathcal{CH}_{(\mathrm{i})}$ at PEV bus $i \in \mathcal{I}_{PEV}$, $\mathcal{I}_{PEV}$ is the set of busses with PEV charger connections, $\mathcal{CH}_{(\mathrm{i})}$ is the set of chargers connected to bus i, and γ is the decision variable vector. The PEV and DG voltage support depends mainly on γ, which can generally take the following form:

$$\gamma = \left[\mathbb{X}_{\left(ch_{(i)},t\right)}, Q^{PEV}_{o(i,t)}, P_{o(i,t)}, Q_{o(i,t)}\right] \tag{30}$$

where $\mathbb{X}_{\left(ch_{(i)},t\right)}$ and $Q^{PEV}_{o(i,t)}$ are the vector of the charger decisions and PEV reactive power at bus $i \in \mathcal{I}_{PEV}$, respectively; $P_{o(i,t)}$ and $Q_{o(i,t)}$ are the DG active and reactive powers at bus $i \in \mathcal{I}_{DG}$, respectively; and $\mathcal{I}_{DG}$ is the set of busses with DG connections. The charging decisions are continuous, that is, $\mathbb{X} \in [0, 1]$ where "0" stands for no charging and "1" stands for full charging. According to the grid operator, γ can be partially constrained. For instance, the PEV reactive powers can be set to zero, that is, $Q_{o(i,t)}^{PEV} = 0,\ \ \forall i \in \mathcal{I}_{PEV}, t$, when the PEV voltage support is disregarded. Stage (I) should satisfy the power flow constraints, as given by

$$P_{G(i,t)} - P_{L(i,t)} = \sum_{j \in \mathcal{I}_b} V_{(i,t)} V_{(j,t)} Y_{(i,j)} \cos\left(\theta_{(i,j)} + \delta_{(j,t)} - \delta_{(i,t)}\right) \qquad \forall i \in \mathcal{I}_b, t \tag{31}$$

$$Q_{G(i,t)} - Q_{L(i,t)} = \sum_{j \in \mathcal{I}_b} V_{(i,t)} V_{(j,t)} Y_{(i,j)} \sin\left(\theta_{(i,j)} + \delta_{(j,t)} - \delta_{(i,t)}\right) \qquad \forall i \in \mathcal{I}_b, t \tag{32}$$

where $P_{G(i,t)}$ and $Q_{G(i,t)}$ denote the generated active and reactive powers, respectively; $P_{L(i,t)}$ and $Q_{L(i,t)}$ are the active and reactive power demands, respectively; $V_{(i,t)}$ and $\delta_{(i,t)}$denote the magnitude and angle of the voltage, respectively; $\mathcal{I}_b$ is the set of system busses; and $Y_{(i,j)}$ and $\theta_{(i,j)}$ are the magnitude and angle of the Y-bus admittance matrix, respectively.

The voltage and feeder thermal limits should also hold, and thus,

$$V_{\min} \le V_{(i,t)} \le V_{\max}, \quad \forall i \in \mathcal{I}_b, t \tag{33}$$

$$I_{(l,t)} \le I_{(l)}^{CAP}, \quad \forall l \in \mathcal{L}, t \tag{34}$$

where $V_{\min}$ and $V_{\max}$ are the maximum and minimum voltage limits, that is, 0.9 and 1.1 p.u., respectively; $I_{(l,t)}$ denotes the per-unit current through line $l \in \mathcal{L}$; $\mathcal{L}$ is the set of system lines; and $I_{(l)}^{CAP}$ is the current carrying capacity.

Typically, two back-to-back power electronic converters are used to interface PEVs and PVs, that is, DC/DC and DC/AC converters. The DC/DC converter performs MPPT with PV-based DGs or controls the PEV charging. The DC/AC converter regulates the DC link voltage and is responsible for the reactive power support [7]. The power injected to a bus should be equal to the output power of the DG installed at that bus:

$$\begin{cases} P_{G(i,t)} = P_{o(i,t)} \\ Q_{G(i,t)} = Q_{o(i,t)} \end{cases}, \quad \forall i \in \mathcal{I}_{DG}, t \tag{35}$$

$$P_{o(i,t)} \le P_{o(i,t)}^{MPPT}, \quad \forall i \in \mathcal{I}_{DG}, t \tag{36}$$

where $P_{o(i,t)}^{MPPT}$ denotes the DG maximum power available. In both PEVs and PVs, the DC/AC converter is similar to that used with Typ. 4 wind farms. Therefore, the reactive power capability limits, defined in [17], should be used as constraints. These limits depend on the converter's power rating and DC link voltage, as follows:

$$Q_{o(i,t)}^2 \le S_{o(i,t)}^2 - P_{o(i,t)}^2, \quad \forall i \in \mathcal{I}_{DG}, t \tag{37}$$

$$\left(Q_{o(i,t)} + \frac{V_{(i,t)}^2}{X_{(i)}}\right)^2 \le \left(\frac{V_{c(i)}^{\max} V_{(i,t)}}{X_{(i)}}\right)^2 - P_{o(i,t)}^2, \quad \forall i \in \mathcal{I}_{DG}, t \tag{38}$$

where $V_{c(i)}^{\max}$ represents the maximum converter voltage which depends on the converter DC link voltage [17, 18], $S_{o(i,t)}$ denotes the DG rated power, and $X_{(i)}$ is the total reactance of the DG filter and interfacing transformer at bus i. If the DC/AC converter increases the set point for the DC link voltage to relax Constraint (38), the DC/DC converter will operate at a high duty cycle, which decreases its efficiency [19]. Hence, the reactive power support from the DC/AC converter is limited by the DC link voltage.

The load power at a bus should be equal to the total power consumed by regular loads and PEV:

$$P_{L(i,t)} = P_{NL(i,t)} + P_{o(i,t)}^{PEV}, \qquad \forall i \in \mathcal{I}_b, t \tag{39}$$

$$Q_{L(i,t)} = Q_{NL(i,t)} + Q_{o(i,t)}^{PEV}, \qquad \forall i \in \mathcal{I}_b, t \tag{40}$$

where $P_{o(i,t)}^{PEV}$ is the PEV active power and $P_{NL(i,t)}$ and $Q_{NL(i,t)}$ are the active and reactive powers of normal loads, respectively. The PV power profile relies mainly on solar insolation, whereas $P_{o(i,t)}^{PEV}$ depends on charging decisions $\mathbb{X}_{(ch_{(i)},t)}$, the charging power limit in kW $P_{CH(ch_{(i)},t)}$, and the charging efficiency $\eta_{CH(ch(i))}$, as given by.

$$P_{o(i,t)}^{PEV} = \sum_{ch_{(i)} \in \mathcal{CH}_{(i)}} \frac{\mathbb{X}_{(ch_{(i)},t)} P_{CH(ch_{(i)},t)}}{\eta_{CH(ch_{(i)})} S_{base}}, \qquad \forall i \in \mathcal{I}_{PEV}, t \tag{41}$$

where S_{base} is the base power for the per-unit system in kW. The charging power limit P_{CH} is a function of the PEV battery state of charge (SOC) and is limited by the capacity of the charger, that is, $P_{CH} \le P_{rated}^{Charger}$. This function is dependent on the characteristics of the battery, which can be expressed as

$$P_{CH(ch_{(i)},t)} = f_{(ch_{(i)},t)}\left(SOC_{(ch_{(i)},t)}^{F}\right), \qquad \forall i \in \mathcal{I}_{PEV}, ch_{(i)}, t \tag{42}$$

where $f_{(ch_{(i)},t)}$ is the function that represents the characteristics of the PEV battery and $SOC_{(ch_{(i)},t)}^{F}$ is the reached SOC. The relationship between the energy delivered to a PEV battery and its SOC can be given by

$$E_{D(ch_{(i)},t)} = E_{BAT(ch_{(i)})} \times \frac{\left(SOC_{(ch_{(i)},t)}^{F} - SOC_{(ch_{(i)},t)}^{I}\right)}{100}, \qquad \forall i \in \mathcal{I}_{PEV}, ch_{(i)}, t \tag{43}$$

where $E_{BAT(ch_{(i)})}$ is the battery capacity in kWh and $SOC_{(ch_{(i)},t)}^{I}$ denotes the PEV initial SOC. The SOC of different PEVs are updated according to

$$SOC_{(ch_{(i)},t)}^{F} = SOC_{(ch_{(i)},t)}^{I} + \frac{\mathbb{X}_{(ch_{(i)},t)} P_{CH(ch_{(i)},t)} \left(\frac{\Delta T}{60}\right)}{E_{BAT(ch_{(i)})}}, \qquad \forall i \in \mathcal{I}_{PEV}, ch_{(i)}, t \tag{44}$$

where ΔT is the time step to collect the system data, run the program, and implement the decisions. Similar to DGs, the injected reactive powers from the PEVs should be limited by their converter ratings and DC link voltages, as given by

$$\left(Q_{o(i,t)}^{PEV}\right)^2 \le \left(S_{o(i,t)}^{PEV}\right)^2 - \left(P_{o(i,t)}^{PEV}\right)^2, \qquad \forall i \in \mathcal{I}_{PEV}, t \tag{45}$$

$$\left(Q_{o(i,t)}^{PEV} + \frac{V_{(i,t)}^2}{X_{(i)}}\right)^2 \le \left(\frac{V_{c(i)}^{\max} V_{(i,t)}}{X_{(i)}}\right)^2 - \left(P_{o(i,t)}^{PEV}\right)^2, \qquad \forall i \in \mathcal{I}_{PEV}, t \tag{46}$$

where $S_{o(i,t)}^{PEV}$ is the rated power of the PEV converter. In addition, the final achieved SOC, that is, $SOC_{(ch_{(i)},t)}^F$, should be limited by the SOC desired by the PEV owners $SOC_{(ch_{(i)},t)}^D$:

$$SOC_{(ch_{(i)},t)}^F \le SOC_{(ch_{(i)},t)}^D, \quad \forall i \in \mathcal{I}_{PEV}, ch_{(i)}, t \tag{47}$$

3.3.2 Problem formulation of stage (II)

In Stage (II), the objective is to minimize the DG active power curtailment, where the final SOC reached in Stage (I), that is, $SOC_{(ch_{(i)},t)}^R$, must be attained to ensure maximum customer satisfaction, which is the highest priority of the V2GQ technology. Therefore, this stage is subject to all of the constraints in Stage (I) except for (47), which is replaced by

$$SOC_{(ch_{(i)},t)}^F = SOC_{(ch_{(i)},t)}^R, \quad \forall i \in \mathcal{I}_{PEV}, ch_{(i)}, t \tag{48}$$

Thus, the objective function of Stage (II) is

$$\max_{\gamma} \sum_{i \in \mathcal{I}_{DG}} P_{o(i,t)}, \qquad \forall t \tag{49}$$

3.3.3 Problem formulation of stage (III)

Stage (III) aims at minimizing the voltage deviation using the DGs and PEVs to restore a feasible solution for the COC and relax the tap operation. Thus,

$$\min_{\gamma} \sum_{i \in \mathcal{I}_b} \left(1 - V_{(i,t)}\right)^2, \quad \forall i, t \tag{50}$$

Besides all the constraints in Stage (II), this problem is subject to the constraint defined in (50), in which the maximum injected powers from the DGs reached in Stage (II), that is, $P_{o(i,t)}^R$, should be maintained.

$$P_{o(i,t)} = P_{o(i,t)}^R, \quad \forall i \in \mathcal{I}_{DG}, t \tag{51}$$

3.4 Coordination between V2GQ and COC

The charging decisions and active/reactive dispatch signals produced in Stage (III) are sent to all PEV parking lots and DGs, as shown in **Figure 9**. To ensure that the PEV and DG converters settle at the desired active and reactive power references, a time delay Δt_{conv} is introduced. The converter settling time may vary from 50 to 100 ms, depending on the primary controllers of the DC/AC converter [7]. For slow automatic interactions, such as voltage regulation, the maximum communication time delay is 100 ms as per the IEC 61850 [20]. Thus, Δt_{conv} is assumed to be

200 ms (100 ms for the converter settling time + 100 ms for the communication latency). Lastly, the COC refines the output solution from Stage (III) to ensure that both $V_{\min}^{sys}$ and $V_{\max}^{sys}$ are within the standard voltage limits. The total execution time of the coordination algorithmΔT is 5 minutes.

3.5 Real-time results

In this section, various case studies are presented to validate the robustness and effectiveness of the optimal coordinated voltage regulation algorithm. The 38-bus 12.66-kV distribution system is used as a test system, as shown in **Figure 10**. The system data can be found in [21]. The system is modified to accommodate two PEV parking lots and four PV-based DGs, with power ratings as given in **Figure 10**. The power demands of the two parking lots are extracted from data provided by the TPA for a weekday in 2013. Both parking lots are commercial, where P1 represents a lot in the vicinity of a train station and P2 is a lot located near downtown Toronto. The total number of PEVs during a day is displayed in **Figure 11**. Due to confidentiality, the addresses of the real parking lots are not mentioned. The central control unit receives the desired SOCs and sends the charging decisions to all vehicles in the parking lots. An OPAL real-time simulator (RTS) is used to model the visual test network using the SimPowerSystems blockset, which is available in Simulink/Matlab, and an ARTEMiS plug-in [22]. The network, PEV, and DG models are

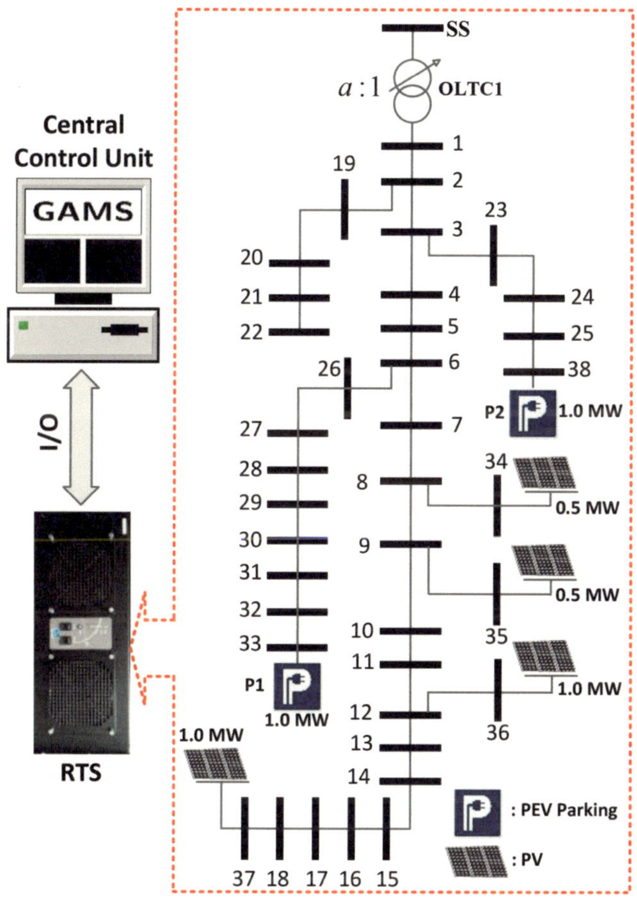

Figure 10.
Test network with an HiL realization.

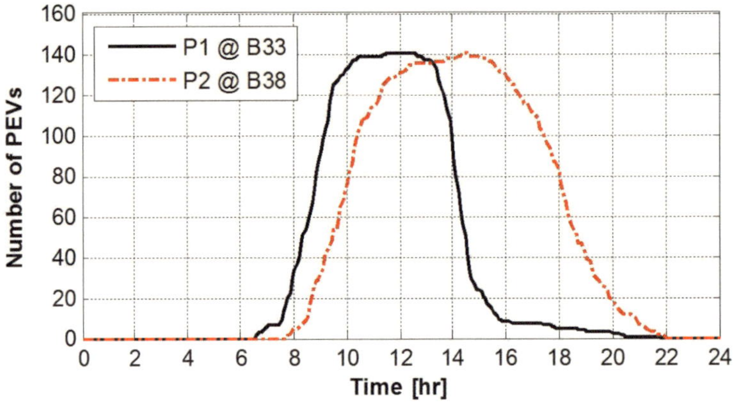

Figure 11.
Number of vehicles in the parking lots.

distributed between the RTS cores for performing parallel computations. The RTS is used to perform a hardware-in-the-loop (HiL) realization, where a central control unit, emulated by a host computer running GAMs, exchanges real-time data with the test network modeled in the RTS. The sampling time used to realize the HiL application is 100 μs.

3.5.1 OLTC control without PEV/DG voltage support

This section compares the responses of the conventional and COC controllers for the OLTC. The voltage support from PEVs and DGs is disabled to study their

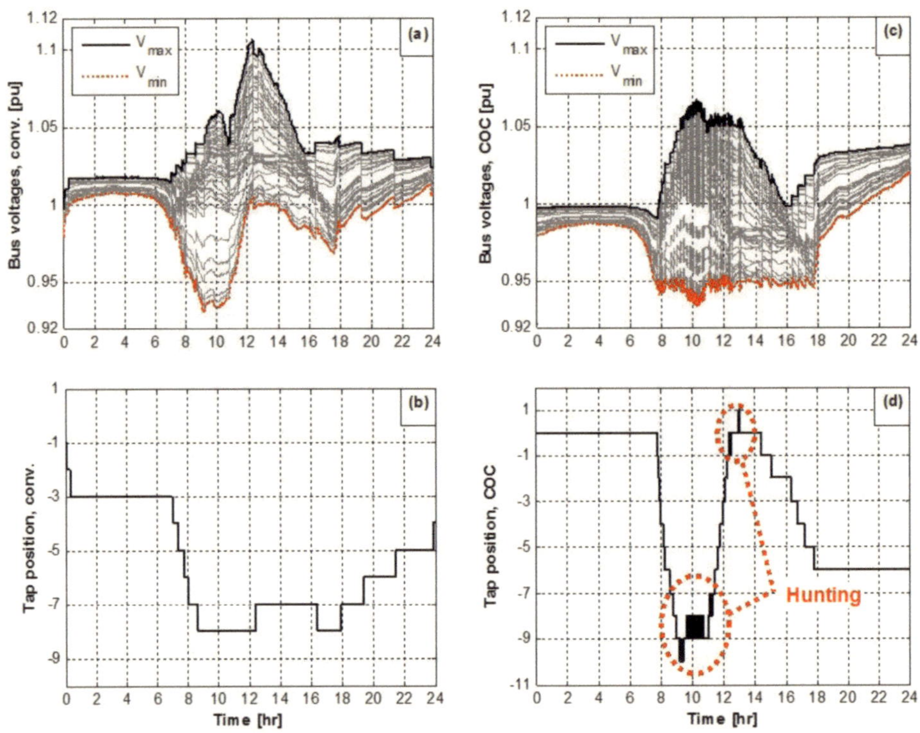

Figure 12.
OLTC response: (a & b) conventional control and (c & d) COC.

impacts of the OLTC. **Figure 12(a)** and **(b)** illustrate the response of the conventional controller for the OLTC over 24 hours. Although the OLTC does not suffer from an excessive tap operation (13 taps/day), undervoltage and overvoltage problems occur.

As expected, the overvoltage happens during the peak power generation from DGs, while the undervoltage coincides with the peak demand of PEV. The COC is enabled to mitigate these voltage violations. **Figure 12(c)** and **(d)** demonstrate that the COC can limit the voltage violations without the PEV/DG voltage support, but it suffers from a hunting problem. This problem happens when the overvoltage and undervoltage occur concurrently. In this situation, the COC should be deactivated.

3.5.2 OLTC control with PEV/DG voltage support

To address the hunting problem, presented in the previous case, the V2GQ is coordinated with the COC. Two case studies dealing with PEV/DG voltage support are carried out, as follows:

1. DG active power curtailment, without PEV and DG reactive power supports, that is, $\gamma = \left[\mathbb{X}_{(ch_{(i)},t)}, P_{o(i,t)}\right]$, where $Q^{PEV}_{o(i,t)} = 0, \forall i \in \mathcal{I}_{PEV}$ and $Q_{o(i,t)} = 0, \forall i \in \mathcal{I}_{DG}$

2. PEV and DG reactive power dispatch, that is, $\gamma = \left[\mathbb{X}_{(ch_{(i)},t)}, Q^{PEV}_{o(i,t)}, P_{o(i,t)}, Q_{o(i,t)}\right]$

Figure 13 demonstrates the performance of the coordination algorithm when the voltage support is merely achieved via the DG active power curtailment (i.e., the

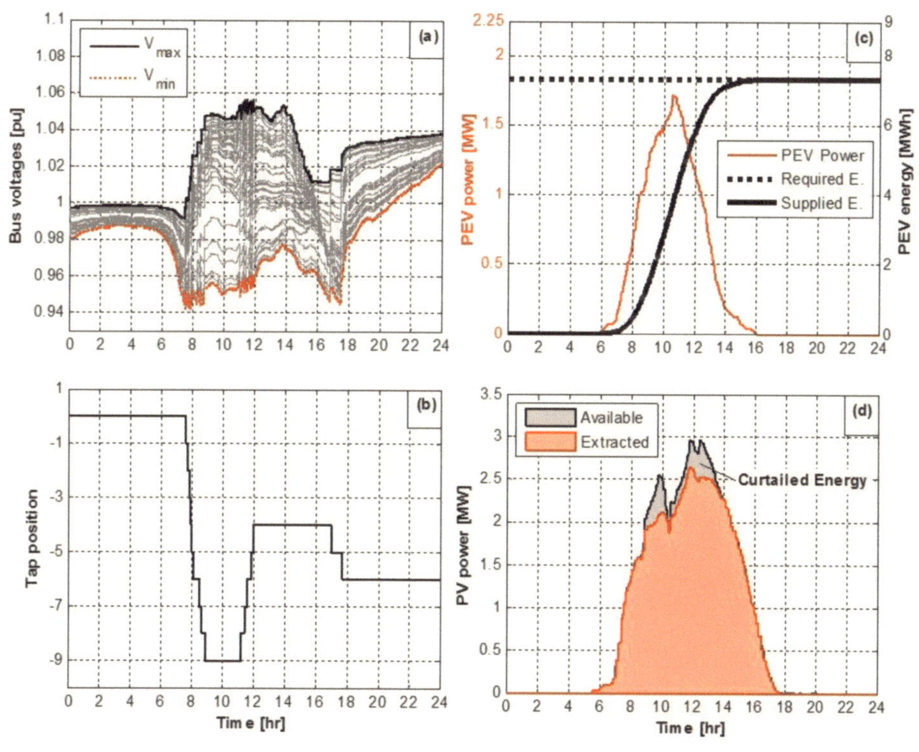

Figure 13.
Response of coordination algorithm, assuming DG active power curtailment.

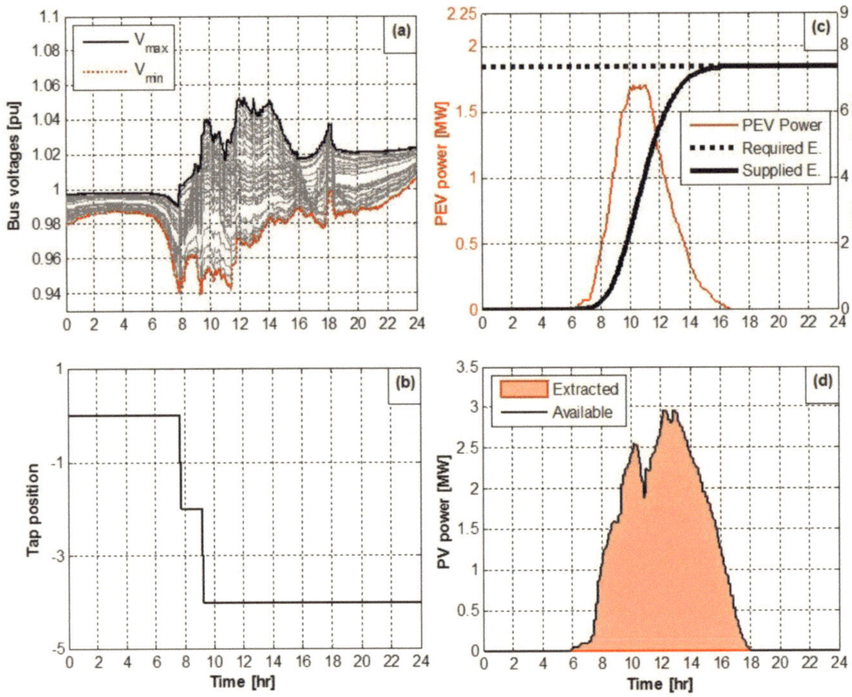

Figure 14.
Response of coordination algorithm, activating both PEV and DG reactive power dispatch.

first case study). The coordination algorithm keeps the bus voltages within the standard limits, as shown in **Figure 13(a)**. The hunting problem is avoided with a reasonable daily tap operation, that is, 16 taps/day; see **Figure 13(b)**. Further, the required PEV charging demand is satisfied, as illustrated in **Figure 13(c)**. However, 6.14% of the DG available energy is curtailed because the priority is given to supplying the PEV demand, as shown in **Figure 13(d)**. This privilege is considered to comply with the distribution system code developed by the Ontario Energy Board [23]. It states that electric utilities should deliver the required energy to supply their loads (such as PEVs) unless there is a technical limit violation. The only solution to maximize the energy extraction is therefore to incorporate both the PEVs and DGs in the voltage support. **Figure 14** illustrates the response of the coordination algorithm for the second case when PEV and DG reactive powers are employed for voltage regulation. Utilizing the full features of the V2GQ results in a proper voltage regulation using only 4 taps/day, extracting all DG power and charging all PEVs.

4. Conclusion

This chapter explained the conventional voltage regulation schemes used in smart grids with inverter-based DGs and PEVs. High penetration levels of PEVs and DGs may lead to negative impacts on the conventional voltage control devices such as tap-changing transformers and capacitor banks. The main reason behind these negative impacts is the stochastic power profiles of PEVs and renewable-based DGs that stimulate chronological overvoltage and undervoltage and make load forecasting erroneous. In the literature, there are two approaches to mitigating the voltage violations associated with DGs and PEVs: (i) centralized and (ii) distributed voltage regulation schemes. These schemes necessitate communication and, thus, may benefit from the communication infrastructure embedded within smart grids. The centralized approach employs state estimation and solves an optimization problem to dispatch DG reactive power for optimal voltage regulation. On the other hand, the distributed approach is an expert-based control or model-free approach, which coordinates a variety of voltage control devices with the goal of providing effective and nonoptimal voltage regulation with fewer communication requirements. Case studies for a centralized voltage control scheme illuminated the role of PEV and DG reactive powers in providing optimal voltage regulation with relaxed tap operation.

Author details

Maher Azzouz
University of Windsor, Windsor, Ontario, Canada

*Address all correspondence to: mazzouz@uwindsor.ca

References

[1] Estimating the Costs and Benefits of the Smart Grid: A Preliminary Estimate of the Investment Requirements and the Resultant Benefits of a Fully Functioning Smart Grid. Palo Alto, CA: EPRI; 2011

[2] The Smart Grid: An Estimation of the Energy and CO2 Benefits. Pacific Northwest National Lab; 2010

[3] Vovos PN, Kiprakis AE, Wallace AR, Harrison GP. Centralized and distributed voltage control: Impact on distributed generation penetration. IEEE Transactions on Power Apparatus and Systems. 2007;**22**(1):476-483

[4] Faiz J, Siahkolah B. Differences between conventional and electronic tap-changers and modifications of controller. IEEE Transactions on Power Delivery. 2006;**21**(3):1342-1349

[5] Zamani MA, Yazdani A, Sidhu TS. A control strategy for enhanced operation of inverter-based microgrids under transient disturbances and network faults. IEEE Transactions on Power Delivery. 2012;**27**(4):1737-1747

[6] Karimi-Ghartemani M, Iravani MR. A method for synchronization of power electronic converters in polluted and variable-frequency environments. IEEE Transactions on Power Apparatus and Systems. 2004;**19**(3):1263-1270

[7] Yazdani A, Iravani R. Voltage-Sourced Converters in Power Systems. Wiley-IEEE Press; Feb 2010. ISBN-10: 0470521562

[8] Katiraei F, Iravani R, Hatziargyriou N, Dimeas A. Microgrids management. The IEEE Power & Energy Magazine. 2008;**6**(3):54-65

[9] Azzouz MA, Farag HE, El-Saadany EF. Real-time fuzzy voltage regulation for distribution networks incorporating high penetration of renewable sources. IEEE Systems Journal. Sep 2017;**1**(3): 1702-1711

[10] Daratha N, Das B, Sharma J. Coordination between OLTC and SVC for voltage regulation in unbalanced distribution system distributed generation. IEEE Transactions on Power Apparatus and Systems. 2013;**29**(1):1-11

[11] Agalgaonkar YP, Pal BC, Jabr RA. Distribution voltage control considering the impact of PV generation on tap changers and autonomous regulators. IEEE Transactions on Power Apparatus and Systems. 2014;**29**(1):182-192

[12] Bonfiglio A, Brignone M, Delfino F, Procopio R. Optimal control and operation of grid-connected photovoltaic production units for voltage support in medium-voltage networks. The IEEE Transactions on Sustainable Energy. 2014;**5**(1):254-263

[13] Azzouz MA, Shaaban MF, El-Saadany EF. Real-time optimal voltage regulation for distribution networks incorporating high penetration of PEVs. IEEE Transactions on Power Apparatus and Systems. 2015;**30**(6)

[14] Milano F. Hybrid control model of under load tap changers. IEEE Transactions on Power Delivery. 2011; **26**(4):2837-2844

[15] Elkhatib ME, El-Shatshat R, Salama MMA. Novel coordinated voltage control for smart distribution networks with DG. IEEE Transactions on Smart Grid. 2011;**2**(4):598-605

[16] Farag HE, El-Saadany EF, Seethapathy R. A two ways communication-based distributed control for voltage regulation in smart

distribution feeders. IEEE Transactions on Smart Grid. 2012;**3**(1):271-281

[17] Ullah NR, Bhattacharya K, Thiringer T. Wind farms as reactive power ancillary service providers—Technical and economic issues. IEEE Transactions on Energy Conversion. 2009;**24**(3): 661-672

[18] Zhou K, Wang D. Relationship between space-vector modulation and three-phase carrier-based PWM: A comprehensive analysis [three-phase inverters]. IEEE Transactions on Industrial Electronics. 2002;**49**(1): 186-196

[19] Kazimierczuk MK. Pulse-Width Modulated DC-DC Power Converters. John Wiley & Sons; 2008

[20] Communication Networks and Systems in Substations—Part 5: Communication Requirements for Functions and Device Models. IEC 61850-5 ed2; 2003

[21] Singh D, Misra RK, Singh D. Effect of load models in distributed generation planning. IEEE Transactions on Power Apparatus and Systems. 2007;**22**(4): 2204-2212

[22] Azzouz MA, El-Saadany EF. Multivariable grid admittance identification for impedance stabilization of active distribution networks. IEEE Transactions on Smart Grid. 2016:1-13. in press

[23] Ontario Energy Board. Distribution system code. 2013. Available from: http://www.ontarioenergyboard.ca/OEB/Documents/Regulatory/DistributionSystemCode.pdf

Chapter 5

A Reliable Communication Model Based on IEEE802.15.4 for WSANs in Smart Grids

Jafar Rasouli, Ahmad Motamedi, Mohamad Baseri and Mahshad Parsa

Abstract

Creating cyber-physical systems (CPSs) based on wireless sensor and actuator networks (WSANs) has great potential to improve the performance of Smart Grid. In addition, IEEE802.15.4 has widely been regarded as an appropriate standard for WSANs, due to some striking and unique features. WSANs require provisioning strict quality of service (QoS) due to noisy harsh environments in Smart Grid applications. Although analytical models have been studied in the literature, they have not provided a full-fledged model for Smart Grid. In this paper, we have added a MAC-level buffer, and a novel Markov chain model has been also proposed. By comparison with previous studies, retransmission confines, acknowledgment, packet length variation, saturated traffic, and degenerate distribution of packet generation are accounted for. The algorithm has been experimentally implemented and appraised on a platform with self-designed WSAN. The analytical model predicts well our exhaustive experiments. Further, Monte Carlo simulations validate mathematical results.

Keywords: smart grid, wireless sensor and actuator networks, IEEE 802.15.4, Markov chain, periodic

1. Introduction

Cyber-physical systems (CPSs) are generally defined as integration of computing and communications technology in order to take control of physical elements [1]. Smart Grid, as an example of a CPS, is a modern power infrastructure to enhance efficiency, reliability, and security [2] along with stable renewable energy production and alternative energy resources. Smart Grid has been designed and implemented through modern communication technologies and automatic control systems [3–5]. Establishing such a complex and elaborate system needs the contribution of sundry technologies. Moreover, everyday life and power networks are inextricably intertwined nowadays. In that every failure, even small, imposes skyrocketing economic and human costs. Therefore, designing a stable and reliable system appears inevitable.

Recently, wireless sensor and actuator network (WSAN) applications have entered a new era of CPS developments like cyber transport system (CTS) and specifically Smart Grid as our research focus [6]. There are various factors which conspicuously impact upon the performance of Smart Grid. Among all the factors, online monitoring and reacting have great capabilities in improving reliability which can be provided by WSANs in every part of Smart Grid assemblage, from generation to consumption [7, 8].

Due to WSANs' low costs, they can specifically affect the distributed generation and the production of renewable energy in generation part. Moreover, posts, overhead, and underground transmission lines are better to be online monitored through WSANs in transmission and distribution part. Eventually, WSANs can be employed in consumption part for substation and residential distribution networks, especially smart meters (AMI) [9] which are shown in **Figure 1**. Although WSANs provide numerous advantages, they encounter some challenges in issues such as real-time data delivery and high-rate data generation, since they have not been specifically designed for Smart Grid.

As often as not, WSAN applications utilize IEEE802.15.4 which take advantage of a low-power link leading to a low data rate transmission (250 kb/s) [10].

Impulsive and robust noises of a power system environment and IEEE802.15.4 standard's intrinsic challenges force us to provide a minimum quality of service (QoS) level to control and monitor applications of Smart Grid [11].

The delay and reliability of WSAN are two prominent parameters in Smart Grid. In order to reach a required QoS level through optimizing network parameters, an elaborate analytical model is essential which is substantially similar to the reality. **Table 1** depicts a summary of the most important applications in Smart Grid and their QoS levels in terms of data rate, latency, and reliability [2, 12]. As the table shows, in contrast to the reliabilities, the range of delays are relatively high.

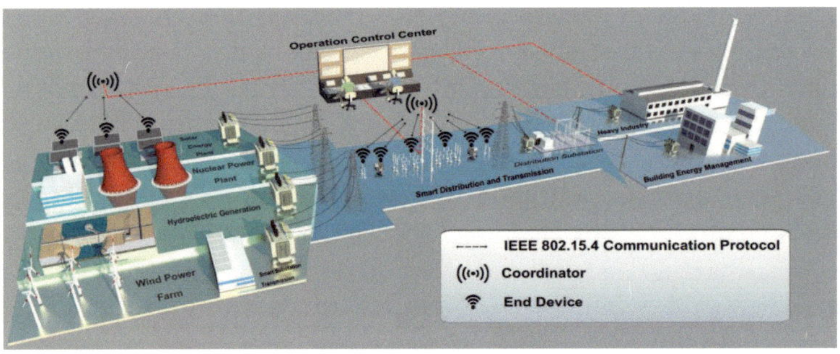

Figure 1.
Smart grid ecosystem monitoring and controlling via WSAN.

Reliability	Latency	Bandwidth (Kbps)	Application
9–99.99%	2–15 s	10–100	AMI[a]
99–99.99%	500 ms–several min	14–100	Demand response
99–99.99%	20 ms–15 s	9.6–56	Distribution energy Resources and storage
99–99.999%	100 ms–2 s	9.6–100	Distribution grid management
99–99.99%	2 sec–5 min	9.6–56, 100 is a good target	Electric Transportation
99–99.99%	2–15 s	10–100	AMI Demand response

[a]*Advanced metering infrastructure.*

Table 1.
Communication requirements of smart grid technologies.

In order to describe the performance of WSANs, several analytical models are introduced in the literature. Some of these models are complicated enough not to be able to be implemented. On the other hand, some others suffer from their low precision due to simplifying and ignoring some parameters such as retransmission and buffer.

Most significantly, a vast majority of the models reach a consensus on using Poisson traffic pattern as the distribution of network input traffic [13–17]. However, delving further into the issue reveals that applications like remote monitoring and Smart Grid generate data with deterministic distribution. In other words, in these applications, each node produces data in a periodic pattern. To illustrate the concept, consider an AMI connected in a consumer side for monitoring and controlling. Based on AMI type and its protocol, the node sends data to the control center every second or minute which this fact shows that AMI data generation is periodic [18–23].

The main contribution of this paper is designing a novel analytical model for IEEE802.15.4 standard. The proposed model is specifically appropriate for applications in which the data is periodically generated such as in industry applications and Smart Grid. In these applications, on the one hand, packets are being produced based on a certain periodic time pattern. On the other hand, service time is always a random variable with general distribution. Therefore, service time might temporarily exceed the period time which, as an inevitable consequence, some packets might encounter a busy channel and be dropped. We solve this problem by proposing our MAC-level queue. We demonstrate that the proposed MAC-level queue not only increases the throughput, but also the direct connection between the generation (sensors) and communication packet systems is eliminated which makes the system far more stable.

Moreover, in order to enhance the proposed model, we have employed retransmission scheme, variable packet length, and saturated traffic condition.

1.1 Cybersecurity

As stated by the Electric Power Research Institute (EPRI), one of the most challenges facing Smart Grid deployment is related to cyber security, and due to the increasing potential of cyberattacks and incidents against this critical sector, it becomes more and more interconnected. A large part of research of many organizations working on the development of Smart Grid such as NIST, NERC-CIP, ISA, IEEE 1402, and NIPP are devoted to security programs. In this paper we suggested a well-known standard, IEEE802.15.4, which the wireless link will be secured in different layers. For example, regarding secure communications, the MAC sublayer offers facilities which can be harnessed by upper layers to achieve the desired level of security. Higher-layer processes may specify keys to perform symmetric cryptography to protect the payload and restrict it to nodes or just a point-to-point link; these nodes can be specified in access control lists. Furthermore, MAC computes freshness checks between successive receptions to ensure that presumably old frames, or data which is no longer considered valid, do not transcend to higher layers. In addition, there is another insecure MAC mode, which allows access control lists merely as a means to decide on the acceptance of frames according to their (presumed) source.

The rest of the paper is organized as follows. In Section 2, we summarize related work. Section 3 lists the main contributions of the paper and their relation with literature. In this section we proposed an extended Markov. Reliability is analyzed accurately in Section 4. In addition, an accurate analysis of packet service time and end-to-end delay is investigated in Section 5. Numerical and simulation results are presented in Section 6. In this section we validate our analysis by experimental results and Monte Carlo simulations. Finally, Section 7 concludes the paper.

2. Related works

The performance of the MAC sublayer in IEEE802.15.4 standard has been evaluated in the literature. The crowd of earlier investigations was based upon MAC sublayer simulation. Lu et al., [24] and Zheng and Lee [25] performed their research based upon simulation.

Gradually, analytical models emerged in this research area where Cao et al., [26] presented an analytical model which was only able to calculate the throughput. Some other models were only able to calculate the energy like [27]. Furthermore, with the passage of time, Markov chain analytical models were proposed, the majority of which are based on [28]'s results.

It should be also pointed out that Bianchi's model [28] was not proper for IEEE802.15.4 standard due to the different functionalities of CSMA/CA mechanism in IEEE 802.11 and IEEE802.15.4 standards.

In [14], although the authors presented an analytical Markov model to evaluate MAC sublayer in the presence of uplink and downlink saturated traffic, the model suffers from the lack of retransmission.

In 2009, despite Yung's efforts to consider retransmission in their proposed model, packet length and acknowledgment were ignored.

In more developed models, Faridi et al., [29] employed retransmission, packet length, and acknowledgment in their advanced model.

In [30], a Markov model is provided to evaluate MAC sublayer and calculate the delay, energy, and throughput which suffers from some drawbacks. Not only did they assume unsaturated data, but they also considered predetermined length for the idle state. In our work, we demonstrate that the duration of the idle states depends on the instantaneous network conditions which might obviously change by passing of time.

Owing to this point, we have considered a variable duration of idle states in our proposed model to deal with the changes in network condition.

In [31], the authors used a model focused on CAP (contention access period), to calculate the throughput and energy and evaluate the effects of a finite length buffer on network performance.

In spite of some drawbacks such as the lack of any queues and some problems in the modeling of the idle states, Park [16, 32] developed the model proposed in [14] through adding retransmission in several investigations. In effect, in Park's model, before passing the whole period of idle states' duration, no node is allowed to leave the idle state, when a new packet is generated. In addition, Park [32] used a backoff with duration of 305 μs instead of the 320 μs, which leads to inaccuracy in his experimental tests. In our experiment, a 1 MHz hardware timer is utilized, to enhance timer resolution up to 1 μs and applying 320 μs to aUnitBackoffPeriod.

In [33], the authors provided different services in Smart Grid by introducing of delay-responsive cross layer (DRX) and also prioritizing input data. DRX classifies information into two categories in the application layer. Potential delay is calculated for every input packet regarding the network history. Then, the best decision was taken at the MAC sublayer to achieve minimum delay to send the packet.

DOI: http://dx.doi.org/10.5772/intechopen.84288

3. The analytical model

In this section, an accurate analytical model is proposed for industrial applications as well as Smart Grids.

IEEE802.15.4 specifies physical and MAC layers, a low-rate and low-energy consumption solution [10]. This standard provides two channel access types: slotted CSMA/CA and unslotted CSMA/CA [34, 35]. Further information concerning the standard for enthusiastic readers is in [36–38].

In order not to get involved in useless elaborate calculations, we consider a star topology with a PAN coordinator, N nodes, and the slotted beacon-enabled CSMA/CA mechanism. Acknowledgment is enabled, and a MAC sublayer buffer has also been designed. The input traffic can be saturated, but its distribution is deterministic. We also assume that the arrival rates for all nodes are the same, and nodes start sensing the medium independently. **Table 2** shows the summary of notations we use in our equations and diagrams.

Symbol	Definition
$b_{i,j,k}$	State probability
W_i	Maximum number of random backoffs in stage i
m	*MacMaxCSMABackoffs*
m_0	*MacMinBE*
n	*MacMaxFrameRetries*
α	Probability that channel is busy in CCA1
β	Probability that channel is busy in CCA2
τ	Probability of starting CCA1
L_p	Length of data packet
L_s	Duration of successful transmission
L_c	Duration of failure transmission due to collision
L_{ack}	Duration of acknowledgment packet
$L_{w,ack}$	Acknowledgment waiting time
$L_{m,ack}$	Acknowledgment maximum waiting time (ACK time out)
P_c	Probability of collision
λ	Packet arrival rate at the MAC sublayer
ρ	Applied load to the queue
$\overline{T}_{Service}$	Average service time for uplink data block
$\overline{T}_{backoff,i}$	Average service time for backoff stage
$\overline{T}_{CCA,i}$	Average service time of carrier sensing
W	Packet waiting time in the queue
W_0	The mean remaining service time
Q	Number of packet in the queue

Table 2.
Summary of notations.

3.1 Deterministic traffic model

In this section, we present a novel model inspired by the Park's model in [16]. There are several deficiencies in Park's model where the author has not considered any buffers; consequently, it was not an accurate and appropriate model for saturated networks. Another problem is that a node remained in idle states for a definite period of L_0, even if a new packet is ready for transmission. Despite of a solution proposed in [33], it is not thorough enough to solve the problem precisely. By modifying the queue as well as idle states and obviating the aforementioned downsides, we have accomplished a perfect model, namely, deterministic traffic model (DTM), illustrated in **Figure 2**.

DTM provides two main blocks: transmission and waiting blocks. Packets can experience success or failure in the transmission block. Failure of packets occurs on the account of channel access failure or lack of receiving acknowledgment. The possibility of every event depends on various parameters, such as the number of nodes in the network, packet length, data generation time period, and MAC parameters.

In our proposed model, waiting block, including idle and queue states, has been added to resolve the weak points of the previous models.

As mentioned before, a wide range of models has been designed based on Poisson traffic distribution. While in monitoring applications like Smart Grid, data are generated in a deterministic manner, persuading us to develop DTM.

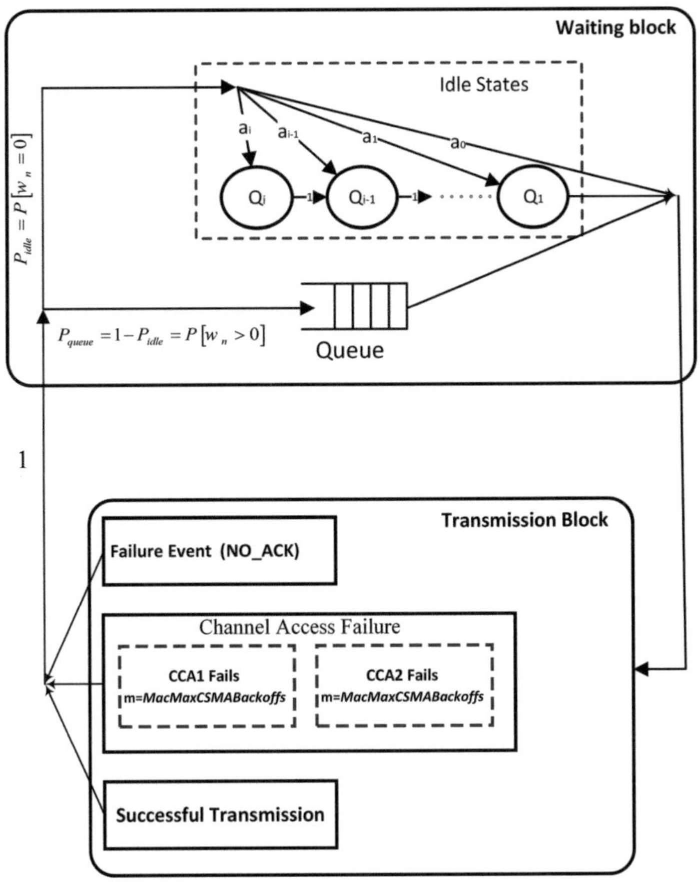

Figure 2.
Proposed deterministic traffic model (DTM).

3.1.1 Waiting block

In the proposed waiting block, idle states consist of variable number of states. The number of the states is determined by the service time, data generation period, and the previous status of buffer. The details of this model come under scrutiny in the following. The queue system is modeled on the D/G/1 FIFO queue. In Kendall notation, D/G/1 FIFO denotes that data packets are generated through a deterministic distribution, while the service time distribution is general [39, 40].

Monte Carlo algorithm and experimental results demonstrate that the service time distribution consists in the MAC parameters such as *MacMinBE*, *MacMaxCSMABackoffs*, and *MacMaxFrameRetries*, L_p and T_P. Changing these parameters effects a change in the shape of service time distribution, in a way that the resulting distribution is similar to none of common probability distributions. Therefore, to be more precise and albeit complicated, general distribution is considered for service time.

In order to derivate waiting block equations such as the idle mode probabilities and waiting time in the queue, we consider a scenario illustrated in **Figure 3**. Assume a periodic sequence of arriving packets with C_n notation. Its probability density function (pdf) is simple impulsive (with the x-intercept = the time period (T_p) and the y-intercept = 1).

u_n is a (key) random variable, defined as

$$u_n = T_{service,\,n} - T_p \tag{1}$$

In a stable network, the expectation value of u_n needs to be negative. Also consider W_n as waiting time:

$$w_{n+1} = \begin{cases} w_n + u_n & \text{if} \quad w_n + u_n \geq 0 \\ 0 & \text{if} \quad w_n + u_n \leq 0 \end{cases} \tag{2}$$

The term $w_n + u_n$ is the sum of unfinished work (w_n) found by C_n plus the service time ($T_{service,n}$) less than T_p. The negative value of this term represents that

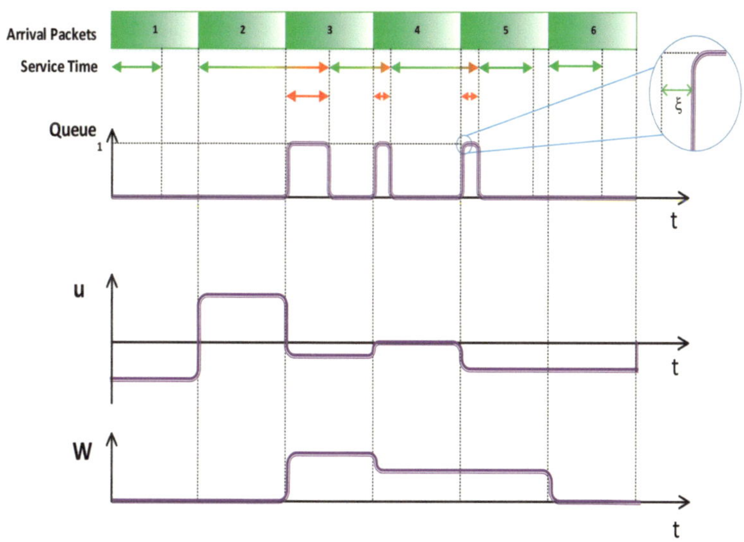

Figure 3.
Time diagram for a scenario, in which a periodic sequence of packets arrives, and queue status, waiting time and the (key) random variable u is shown.

T_p has elapsed since the arrival of C_n and the node must enter idle mode by the time C_{n+1} arrives.

We may write Eq. (2) as

$$w_{n+1} = \max[0, w_n + u_n] \tag{3a}$$

So as to clarify the subject matter, the time diagram for the scenario is illustrated in **Figure 3**. Six packets are generated in determined intervals, and each one takes a different service time to be transmitted. The first packet's service time is less than the time period. But it is more for the second. So, before the second packet departs from its node, the third packet is generated and enters the queue directly. Though the third packet's service time is smaller than T_p, the fourth packet's waiting time is not zero because of the high second packet's unfinished work (w_2) or

$$\sum_{i=2}^{3} (T_{Service,i}) \geq 2T_p. \tag{3b}$$

This may affect several subsequent packets when $w = 0$ for a packet. We define W(y) as cumulative distribution function (CDF) for w_n:

$$W(y) = \lim_{n\to\infty} P[w_n \leq y] \tag{4}$$

Before proceeding with the theory, to calculate $W(y)$, C_n (u) is defined as the CDF for random variable u_n:

$$C(u) = P[u_n \leq u] = \int_{t=0}^{\infty} B(u+t)d\gamma(t - T_p) \tag{5}$$

In which $B(x)$ and $\delta(t\text{-}T_p)$ are the distributions of service time and time period, respectively.

Combining Eqs. (2), (4), and (5), we have Lindley's integral Equation [41] which is seen to be a Wiener-Hopf-type integral Equation [42]:

$$W(y) = \begin{cases} \int_{0^-}^{\infty} C(y-w)dW(w) & y \geq 0 \\ 0 & y < 0 \end{cases} \tag{6}$$

The node must go to idle mode if waiting time equals zero, as illustrated in **Figure 2**:

$$P_{idle} = P[w_n = 0] \tag{7}$$

And the probability of existing at least one packet in the queue is

$$P_{queue} = 1 - P_{idle} = P[w_n > 0] \tag{8}$$

The mean queue length $\overline{Q}$ can be calculated using Little's theorem:

$$\overline{Q} = \frac{\overline{W}}{T_p} \tag{9}$$

And the mean number of packets in the buffer $\overline{K}$

$$\overline{K} = \rho + \overline{Q} \tag{10}$$

In which

$$\rho = \frac{\overline{T_{Service}}}{T_p} \tag{11}$$

As regards the next packet's arrival time is specified, the time that nodes spend on idle mode is conspicuous. As a result, idle mode constitutes several states in DTM. The number of idle states (i), which represents maximum idle mode's waiting time, is obtained from minimum service time in transmission block:

$$u_{\min} \triangleq T_{service,\min} - T_p \tag{12}$$

The minimum service time and its probability are

$$\text{Min}(ServiceTime) = (m+1) \times aUnitBackoffPeriod \tag{13}$$

$$P_{MinServiceTime} = \frac{\alpha^{m+1}}{2^{\frac{m(m+1)}{2}} {W_0}^{m+1}} \tag{14}$$

Waiting time in idle mode for the next packet will decline if the service time for the current packet rises, until service time and time period become equal. Thus, it is required to derivate PDF of u_n which is obtained by Eq. (5):

$$c_n(u) = \frac{dC_n(u)}{du} \tag{15}$$

The smallest time unit in the DTM is equal to $aUnitBackoffPeriod$, but the packet generation period can take continuous values which may not be divisible by $aUnitBackoffPeriod$; consequently $\tilde{c}(u)$is the normalized value of $c_n(u)$:

$$\tilde{c}(u) = \left\lceil \frac{c(u)}{aUnitBackoffPeriod} \right\rceil \times aUnitBackoffPeriod \tag{16}$$

There is always minor inaccuracy imposed to calculation, with a maximum value of approximately aUnitBackoffPeriod. ξ represents the error in **Figure 3**. Accordingly, the probability of entering idle mode is as follows:

$$a_0 = P[\tilde{c}(u) = 0] \tag{17}$$

$$a_1 = P[\tilde{c}(u) = -aUnitBackoffPeriod] \tag{18}$$

$$a_i = P[\tilde{c}(u) = -i] \tag{19}$$

The maximum time that a node remains in idle mode occurs when the packet does not enter the queue and it is transmitted in minimum possible time (minimum service time) as well:

$$i = \left(\left\lceil \frac{T_P}{aUnitBackoffPeriod} \right\rceil - (m+1) \right) \times aUnitBackoffPeriod \tag{20}$$

The expected number of idle states is

$$E[a] = \left[\frac{-\int_{-i}^{0} uc(u)du}{\int_{-i}^{0} c(u)du \times aUnitBackoffPeriod} \right] \tag{21}$$

And idle states probabilities are

$$\begin{aligned}
I_i &= \left[P\left(Failure^{CCA}\right) + P\left(Failure^{NO_ACK}\right) + P(Success)\right] \times P[w_n = 0] \times a_i \\
I_{i-1} &= \left[P\left(Failure^{CCA}\right) + P\left(Failure^{NO_ACK}\right) + P(Success)\right] \times P[w_n = 0] \times a_{i-1} + I_i \\
I_{i-2} &= \left[P\left(Failure^{CCA}\right) + P\left(Failure^{NO_ACK}\right) + P(Success)\right] \times P[w_n = 0] \times a_{i-2} + I_{i-1} \\
&\;\;\vdots \qquad\qquad \vdots \\
I_1 &= \left[P\left(Failure^{CCA}\right) + P\left(Failure^{NO_ACK}\right) + P(Success)\right] \times P[w_n = 0] \times a_1 + I_2
\end{aligned} \tag{22}$$

And sum of idle states probabilities is

$$\begin{aligned}
\sum_{j=1}^{i} I_j = &\left[P\left(Failure^{CCA}\right) + P\left(Failure^{NO_ACK}\right) + P(Success)\right] \\
&\times P[w_n = 0] \times \sum_{k=1}^{i} ka_k
\end{aligned} \tag{23}$$

Furthermore, there is a MAC-level buffer in waiting block that has not been considered in [16], which is completely separated from the idle mode. If a node generates a packet and also has a packet in transmission block, the new packet is directed toward the buffer until its turn. In other words, when service time becomes far more than the time period, the queue starts to fill. The Markov chain model for a FIFO queue buffer is illustrated in **Figure 4**.

According to **Figure 4**

$$B_0 = P[w_n > 0] \times \left[P\left(Failure^{CCA}\right) + P\left(Failure^{NO\_ACK}\right) + P(Success)\right] \tag{24}$$

The total probability of queue states is

$$\sum_{v=0}^{D_0-1} B_v = D_0 B_0 \tag{25}$$

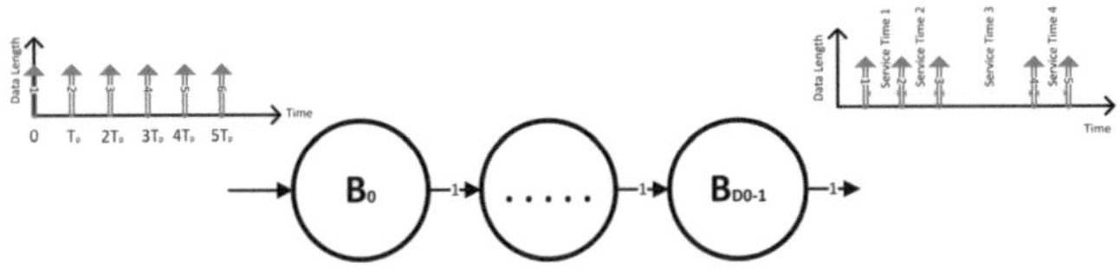

Figure 4.
Markov chain model for MAC-level buffer.

According to Eq. (23) and (25), the total probability of waiting block is

$$\sum P(WaitingBlock) = \sum_{v=0}^{D_0-1} B_v + \sum_{j=1}^{i} I_j \tag{26}$$

3.1.2 Transmission block

The waiting block's probabilities were described in the previous section, and we depict the transmission block details in the following. In this section, some modifications to Park's model [16] are provided. The transmission block accounts for a three-dimensional Markov chain using three stochastic processes, including S(t), backoff stage; *C(t)*, the state of backoff counter; and *r(t)*, the state of retransmission counter. These states are linked to BE, NB, and RT in MAC parameter in IEEE802.15.4 standard, respectively.

The stationary probability of the Markov chain can be written as

$$b_{i,k,j} = \lim_{t\to\infty} P(S(t) = i, C(t) = k, r(t) = j) \tag{27}$$

In which $i\epsilon(-2,m)$, $k\epsilon(-1,max\{W_i\text{-}1,L_s\text{-}1,L_c\text{-}1\})$, and $j\epsilon(0,n)$. **Figure 5** presents Markov chain model for the transmission block. As shown in the figure, the number of retransmissions is considered to have finite values, giving rise to the packet drop. Packets are discarded due to two events:

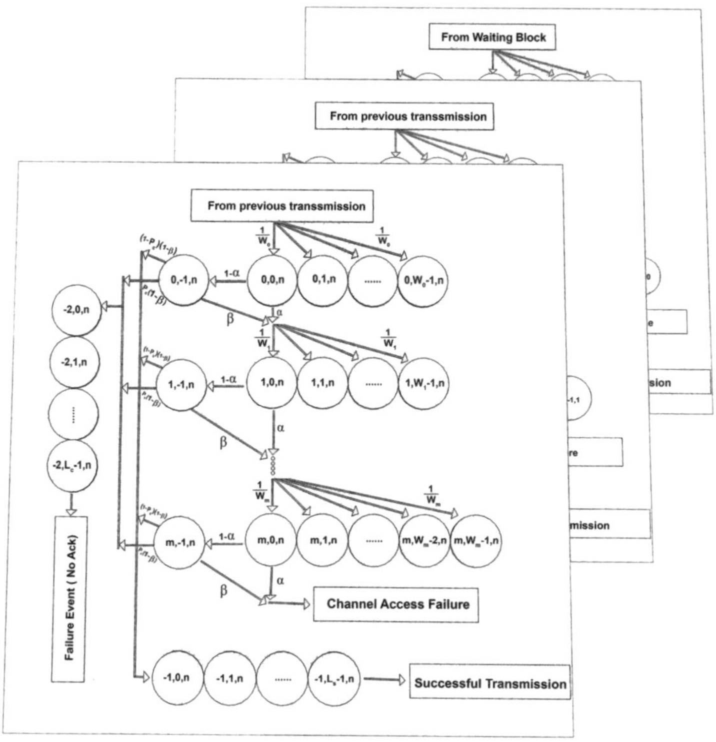

Figure 5.
Markov chain model for MAC-level buffer.

- Channel access failure, working out when one of CCAs within the (m + 1)th backoff fails.

- Lack of receiving acknowledgment (ACK), turning out if collision occurs after n + 1 attempts.

Transmission block features three major parts:

State notations from $(i, W_m\text{-}1, j)$ to $(i, W_0\text{-}1, j)$ denote backoff states. The states $(i, 0, j)$ and $(i, \text{-}1, j)$ are connected to the first and second clear channel assessment (CCA), respectively. States $(-1, k, j)$ and $(-2, k, j)$ correspond to successful and unsuccessful (due to lack of ACK) transmission, respectively. So as to appraise the performance of the network, τ, P_c, α, and β are derived as follows. Interested readers are referred to [16] for more details.

The probability that a device attempts to CCA1 is

$$\tau = \sum_{i=0}^{m}\sum_{j=0}^{n} b_{i,0,j} = \left(\frac{1-x^{m+1}}{1-x}\right)\left(\frac{1-y^{n+1}}{1-y}\right)b_{0,0,0} \tag{28}$$

where $x = \alpha + (1\text{-}\alpha)\beta$ and $y = P_c(1\text{-}x^{m+1})$.

τ depends on α, β, and P_c. The term P_c is the probability that at least one of the $N\text{-}1$ remaining nodes transmits a packet when the channel is occupied:

$$P_c = 1 - (1-\tau)^{N-1} \tag{29}$$

In which N is the number of nodes in the network.

α represents the probability that a node senses the channel and finds it busy in CCA1, (due to data and ACK transmission of other nodes or noise):

$$\alpha = \left[L\left(1-(1-\tau)^{N-1}\right) + L_{ack}\frac{N\tau(1-\tau)^{N-1}}{1-(1-\tau)^{N}}\left(1-(1-\tau)^{N-1}\right)\right](1-\alpha)(1-\beta) \tag{30}$$

When the channel is free in CCA1, it can be busy in CCA2 with the probability of β:

$$\beta = \frac{P_c(1-N\tau)+N\tau}{P_c(1-N\tau)+N\tau+1} \tag{31}$$

In this model

$$\begin{aligned} L_S &= L_P + L_{w,ack} + L_{ack} + L_{IFS} \\ L_c &= L_P + L_{m.ack} \end{aligned} \tag{32}$$

L_s and L_c is the successful transmission time and NO_ACK interval, respectively. L_p represents the total packet length including overhead and payload. $L_{w,ack}$ denotes the ACK waiting time. Lack indicates the length of ACK frame, while LIFS is the interframe spacing (IFS) time, and $L_{m,ack}$ represents ACK packet timeout, determined by *macAckWaitDuration*.

We outline below the final derived transmission block's equations and ignore details:

$$\sum P(Backoff) = \sum_{i=0}^{m}\sum_{k=0}^{W_i-1}\sum_{j=0}^{n} b_{i,j,k} = \frac{1}{2}\left(\frac{1-(2x)^{m+1}}{1-2x}W_0 + \frac{1-(x)^{m+1}}{1-x}\right)\frac{1-(y)^{n+1}}{1-y}b_{0,0,0} \tag{33}$$

Probability of attempting to sense the channel for the second time (CCA2)

$$\sum P(CCA_2) = \sum_{i=0}^{m}\sum_{j=0}^{n} b_{i,-1,j} = (1-\alpha)\frac{1-(x)^{m+1}}{1-x}\frac{1-(y)^{n+1}}{1-y}b_{0,0,0} \tag{34}$$

Finally, the successful and unsuccessful packet transmission (due to NO_ACK)

$$\sum P(Sending) = \sum P(Succ.Trans) + \sum P(Unsucc.Trans) \tag{35}$$

$$\sum P(Succ.Trans) = \sum_{j=0}^{n}\sum_{k=0}^{L_s-1} b_{-1,k,j} = L_s(1-P_c)\left(1-(x)^{m+1}\right)\frac{1-(y)^{n+1}}{1-y}b_{0,0,0} \tag{36}$$

$$\sum P(Unsucc.Trans) = \sum_{j=0}^{n}\sum_{k=0}^{L_c-1} b_{-2,k,j} = L_c P_c\left(1-(x)^{m+1}\right)\frac{1-(y)^{n+1}}{1-y}b_{0,0,0} \tag{37}$$

According to (33), (34), (36), and (37), the total probability transmission block is

$$\begin{aligned}\sum P(TransmissionBlock) &= \sum P(backoff) + \sum P(CCA) + \sum P(Sending) \\ &= \left(\left(L_s(1-P_c) + L_c P_c + \frac{0.5+1-\alpha}{1-x}\right)\right. \\ &\quad \left.\times(1-x^{m+1}) + \frac{1-(2x)^{m+1}}{2(1-2x)}W_0\right) \times \frac{1-y^{n+1}}{1-y}b_{0,0,0}\end{aligned} \tag{38}$$

According to the transmission block's equations, we are in a position to calculate the failure events and successful probabilities for waiting block:

$$P(Success) = \sum_{i=0}^{m}\sum_{j=0}^{n}(1-P_c)(1-\beta)(1-\alpha)b_{i,-1,j} = (1-P_c)\left(1-x^{m+1}\right)\frac{1-y^{n+1}}{1-y}b_{0,0,0} \tag{39}$$

$$P\left(Failure^{NO\_ACK}\right) = \sum_{i=0}^{m} P_c(1-\alpha)(1-\beta)b_{i,0,j} = P_c\left(1-x^{m+1}\right)y^n b_{0,0,0} \tag{40}$$

$$P\left(Failure^{CCA}\right) = \sum_{j=0}^{n}(\alpha+(1-\alpha)\beta)b_{m,0,j} = P_c\left(1-x^{m+1}\right)y^n b_{0,0,0} \tag{41}$$

Obviously speaking, all equations in transmission and waiting blocks depend on $b_{0,0,0}$. So another equation is required to solve these equations. The sum of all states' probability in these two blocks must be equal to one:

$$\sum TransmissionBlock + \sum WaitingBlock = 1 \tag{42}$$

$b_{0,0,0}$ can be calculated by Eq. (42).

Solving nonlinear equations in terms of α, β, and τ leads to find the network quiescent points and also models the behavior of the medium.

In the following, reliability, end-to-end delay, and throughput, as the most critical parameters, are scrutinized.

4. Reliability

The probability of successful packet reception is defined as reliability. In cyber-physical systems, particularly Smart Grid, wireless links may experience a great deal of challenges such as strong noise with heavy-tailed distributions. This means that reliability is a crucial parameter. There are three events in the transmission block, only one of which leads to successful transmission and others are failure events. As mentioned formerly, channel access failure and NO_ACK in the last retransmission are responsible for the failure event.

$$R = 1 - P_{dc} - P_{dr} \tag{43}$$

In which P_{dc} and P_{dr} are the probability of discarded packet (owing to channel access failure) and NO_ACK in the last retransmission, respectively. Following the Markov model illustrated in **Figure 5**

$$P_{dc} = \sum_{j=0}^{n} xb_{m,\,0,j} = \frac{x^{m+1}(1-y^{n+1})}{1-y} \tag{44}$$

$$P_{dr} = P_c(1-\beta)\sum_{i=0}^{m} b_{i,\,-1,\,n} = y^{n+1} \tag{45}$$

5. Packet service time and end-to-end delay

The average delay for successful transmission is defined as the time interval between a packet arrival and the reception of corresponded ACK. It features the waiting time in queue and the service time in the transmission block. In previous works [16, 33], however, the queue delay is overlooked, and the average delay was defined as the time interval from the instant that packet is at the head of its MAC queue until receiving the corresponding ACK. In industrial applications, especially in power grid, delay plays a vitally important role. A delayed command or notification may give rise to chain errors, thereby calculating the precise amount of delay that appears essential.

As mentioned in previous parts, in the transmission process, two consecutive successful CCAs mean that the node is allowed to send its packet. If the node finds the channel busy in each of CCAs, it tries the next backoff stage. This proceeds until m reaches macMaxBE. The service time varies if a node finds the channel busy in each CCA. Attempting to seize the channel for various times is responsible for the different combinations of service time. Let $C_{\alpha\beta}$ *(i)* be the set comprising all the combination of choosing i element out of a set of busy channel probability $S_{\alpha\beta} = \{\alpha, \beta(1-\alpha)\}$. Normalized average service time is as follows:

$$\overline{T}_{CCA,i} = \frac{\sum_{k=1}^{2^i} C_{S_{\alpha\beta}}^k(i)\left(N_\alpha^k(i) + 2N_{\beta(1-\alpha)}^k(i)\right)}{\sum_{k=1}^{2^m} C_{S_{\alpha\beta}}^k(i)} \tag{46}$$

where $C_{S_{\alpha\beta}}^k(i)$ returns the kth combination out of a set of $S_{\alpha\beta}$ in i^{th} backoff attempt. In addition, $N_\alpha^k(i)$ and $N_{\beta(1-\alpha)}^k(i)$ represent the number of the first and second element of set of $S_{\alpha\beta}$ in k^{th} combination, respectively. The delay in backoff stages is presented in (47) (note that the midpoint of the uniform distribution indicates the average value):

$$T_{backoff,i} = \frac{\sum_{k=1}^{2^i} C_{S_{\alpha\beta}}^k(i) \times (W_i - 1)}{2\sum_{k=1}^{2^m} C_{S_{\alpha\beta}}^k(i)} \tag{47}$$

The average time for success and failure transmissions can be calculated according to (46) and (47). The success occurs after j failures (due to NO_ACK):

$$\overline{T}_{success,j} = L_s + 2 + j(L_c + 2) + (j+1)\left[\sum_{i=0}^{m}\left(\overline{T}_{backoff,i} + \overline{T}_{CCA,i}\right)\right] \tag{48}$$

The average time for failures due to limitation of backoff attempt is

$$\begin{aligned}\overline{T}_{Failure,\,CCA} =\ & j(L_c + 2) + j\left[\sum_{i=0}^{m}\left(\overline{T}_{backoff,i} + \overline{T}_{CCA,i}\right)\right] \\ & + \overline{T}_{backoff,m+1} + \overline{T}_{CCA,m+1}\end{aligned} \tag{49}$$

The average time for failures due to the retransmission limit is

$$\overline{T}_{Failure,\,NO\_ACK} = (n+1)\left[L_c + 2 + \sum_{i=0}^{m}\left(\overline{T}_{backoff,i} + \overline{T}_{CCA,i}\right)\right] \tag{50}$$

Accordingly, the average service is

$$\begin{aligned}\overline{T}_{Service} =\ & \sum_{j=0}^{n}\left[P(Success) \times \overline{T}_{success,j} + P\left(Failure^{CCA}\right) \times \overline{T}_{Failure,\,CCA}\right] \\ & + P\left(Failure^{NO\_ACK}\right) \times \overline{T}_{Failure,\,NO\_ACK}\end{aligned} \tag{51}$$

Finally, according to Eqs. (51), (39), (40), and (41), the service time is

$$\begin{aligned}\overline{T}_{Service} =\ & \left[\begin{array}{c}(1-P_c)(1-x^{m+1})\left(L_s + 2 + \sum_{i=0}^{m}\left(\overline{T}_{backoff,i} + \overline{T}_{CCA,i}\right)\right) \\ + x^{m+1}\left(\overline{T}_{backoff,m+1} + \overline{T}_{CCA,m+1}\right)\end{array}\right] \\ & \times \frac{1-y^{n+1}}{1-y} + \left[\left(L_c + 2 + \sum_{i=0}^{m}\left(\overline{T}_{backoff,i} + \overline{T}_{CCA,i}\right)\right)\frac{1-y^{n+1}}{1-y}y\right]\end{aligned} \tag{52}$$

The end-to-end delay consists of the service time and the waiting time:

$$\overline{D} = \overline{T}_{Service} + \overline{W} \tag{53}$$

In which $\overline{T}_{Service}$ is the average service time for the tagged packet and $\overline{W}$ is the waiting time in the queue. The waiting time is made up of the service times for all of packets in the queue ahead of the tagged packet plus the remaining service time of the packet in service (if any).

6. Simulation and analysis

In this section, we have validated our model by drawing a comparison between mathematical results, Monte Carlo simulations, and experimental tests in terms of delay, reliability, and throughput. All experiments are conducted with self-designed motes, each of which features an AT86RF233 amplified ZigBit wireless module and a SAM3S2B microcontroller, both by Atmel. In IEEE802.15.4 standard, aUnitBackoffPeriod is defined as 10 bytes, each of which equals 2 symbols, corresponding to 320 μs in 250 kbps. Ten nodes are positioned in a star topology with beacon-enabled mode. Each node is at the distance of around 20 m from the coordinator, and all nodes are distributed in an area of 1000 m^2.

The impact of the packet generation rate, MAC parameters, and the number of nodes on delay are then evaluated. So as to enhance the reliability of the system, ACK mechanism is activated. The MAC parameters are set according to the standard document [10].

Figure 6 compares the service time, given in the Eq. (52), as a function of various MAC parameters m0, m, and n, obtained from the tagged node (i.e., a node which we perform our evaluations on). The DTM results and Monte Carlo simulations perfectly coincide, and they both predict well the experimental results. As expected, the service time is more sensitive to m0 than m and n. The network becomes unstable, and the buffers are filled if, for a long time, service time is more than 20 ms (horizontal dotted line). Curve fittings are also performed in order to set the optimum parameters in Eqs. (54)–(56):

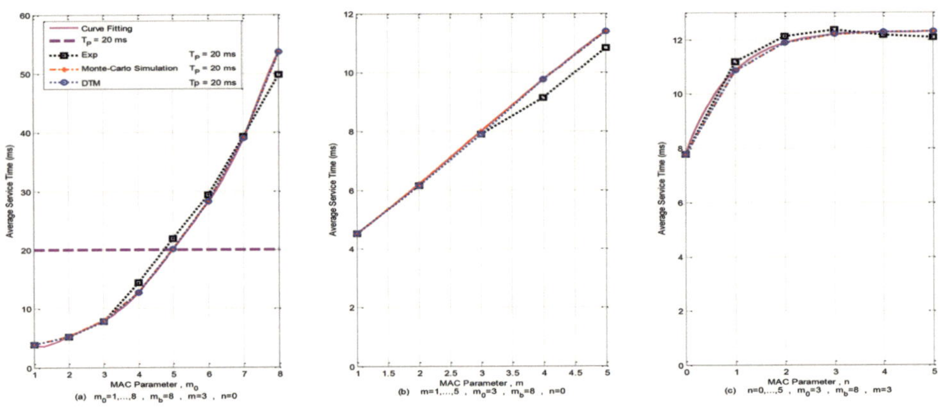

Figure 6.
The average service time as a function of MAC parameters a) m_o = 1, ..., 8, m_b = 8, b) m = 1, ..., 5, n = 0, ..., 5, obtained from DTM, Monte Carlo simulations and experimental tests. The curve fitting of the Monte Carlo simulation for optimization is also added. The length of the packet is L = 2, and the number of nodes is N = 10. Experimental tests are acquired out of 10 runs, each 10^6 aUnitBackoffPeriod.

$$Service\_time(m_0) = -0.00402m_0^7 + 0.12525m_0^6 - 1.5786m_0^5 + 10.367m_0^4 \\ -37.914m_0^3 + 77.214m_0^2 - 78.778m_0^1 + 34.446 \tag{54}$$

$$Service\_time(m) = -0.018055m^4 + 0.19116m^3 \\ -0.65661m^2 + 2.5481m^1 + 2.459 \tag{55}$$

$$Service\_time(n) = 0.005356n^5 - 0.092715n^4 \\ +0.65581n^3 - 2.4454n^2 + 4.9795n^1 + 7.7724 \tag{56}$$

Figure 7 shows the service time's composition as a function of the MAC parameter m and the time period (T_P). When T_P and m are reduced and increased, respectively, the contribution of failure events in the service time will be highlighted. This leads to the reduction of reliability. In T_P = 100 ms, the majority of failure events is due to lack of the ACK packet, but if T_P declines to 10 ms, the channel access failure also appears. In **Figure 7(c)**, raising m up to 4 and 5 makes the network unstable (green dashed line).

Figure 8 illustrates reliability which is obtained by Eq. (43) as a function of m0, m, and n. Like service time, DTM and Monte Carlo simulations perfectly coincide, and both of them predict well the experimental results.

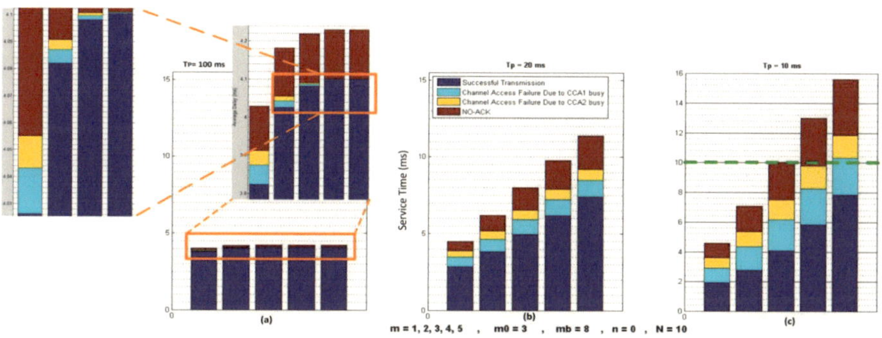

Figure 7.
The service time expected value as a function of MAC parameters m_o = 3, m_b = 8, m = 1, …, 5, n = 0. a) T_p = 100ms b) T_p = 20ms c) T_p = 10ms.

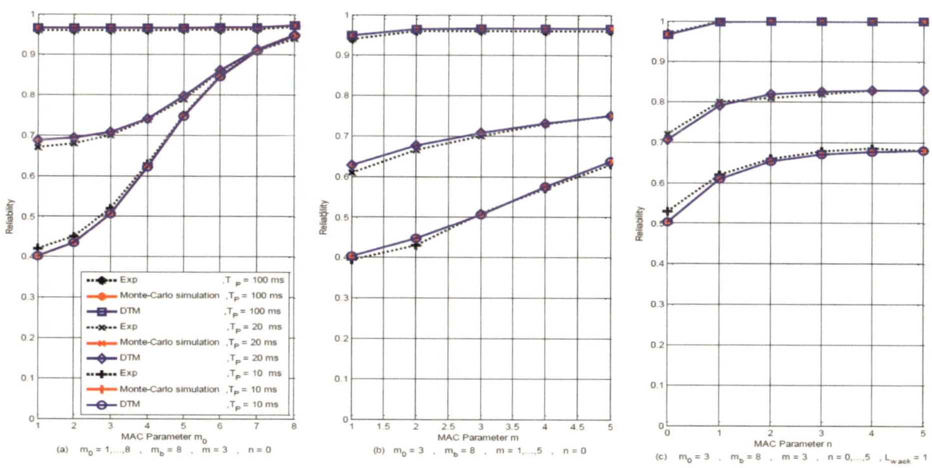

Figure 8.
Reliability as a function of MAC parameters a) m_o = 0, …, 8, m_b = 8, b) m = 1, …, 5, c) n = 0, … 5, obtained by DTM, Monte Carlo simulations, and experimental tests.

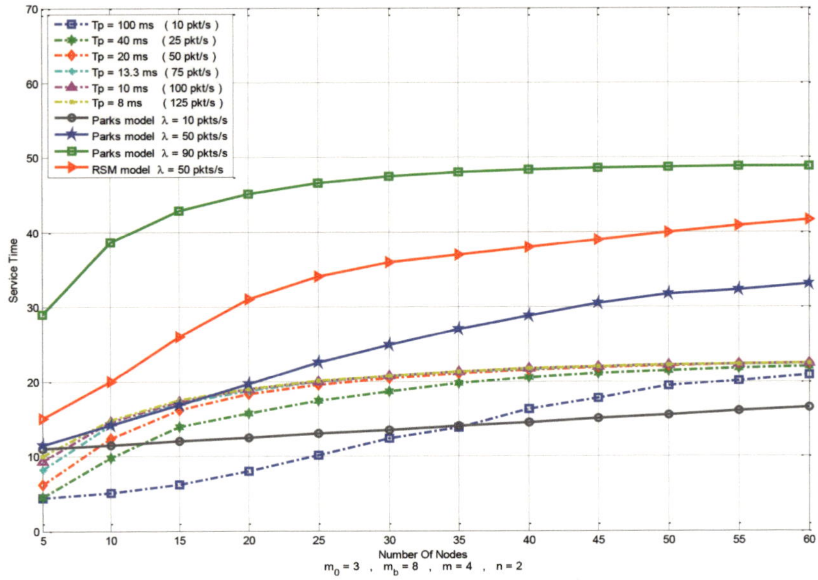

Figure 9.
Average service time against the number of nodes and the data generation period compared to Park's model [32] and RSM model [33]. The length of the packets is L = 2.

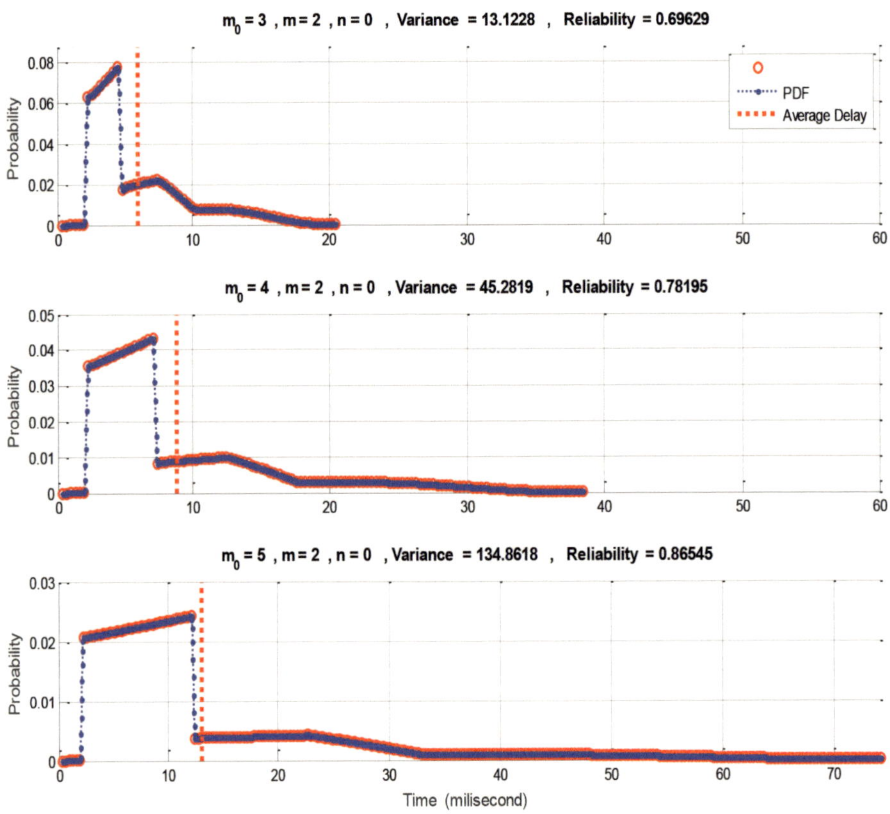

Figure 10.
Average service time against the number of nodes and the data generation period compared to Park's model [32] and RSM model [33]. The length of the packets is L = 2.

Figure 9 depicts the packet transmission service time against the number of nodes for different data generation time periods (note that the waiting time in the queue is ignored in this figure, and it will be evaluated in the following). Changing the input data rate causes large differences in average delay in Park's model [32] (Poisson distribution is considered in Park's model), whereas in DTM model, the average delay does not fluctuate by changing data rate, so it shows a stable behavior which is necessary for Smart Grid. In fact, DTM model is independent of the traffic rate.

Now, the analysis of DTM using probability density function (PDF) of the service time is taken into account. As mentioned in previous sections, whether a node enters the queue or idle states, and also how many states the idle mode has, depends on the PDF's shape. **Figure 10** shows changes in PDF against m0. A rise in

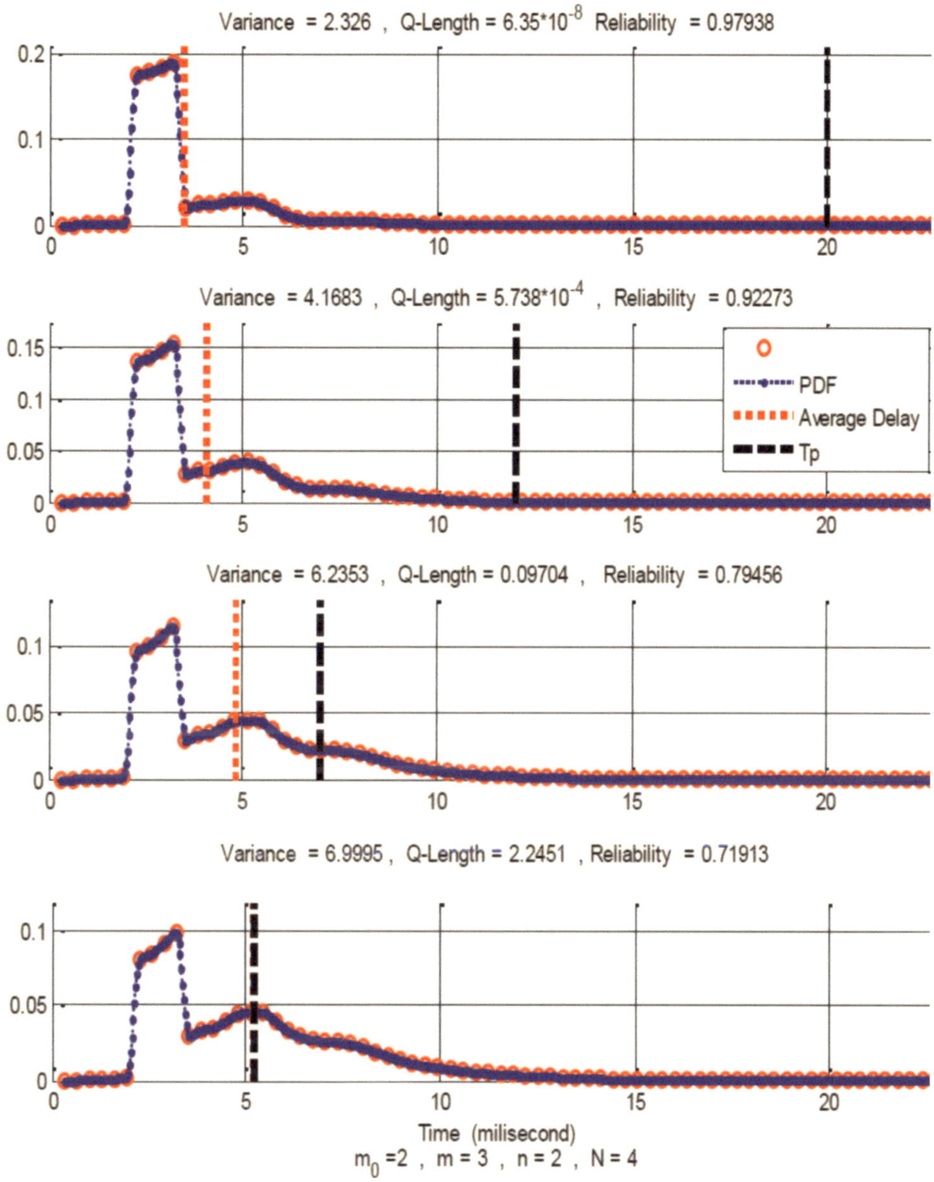

Figure 11.
Service time's PDF against T_P, *while* $m_0 = 2$, $m_b = 8$, $m = 3$, $n = 2$, *and* $N = 4$. *The length of the packet is* L = 2. *Decline in TP contributes to rise in variance, queue length and peak, average delay, and drop in reliability.*

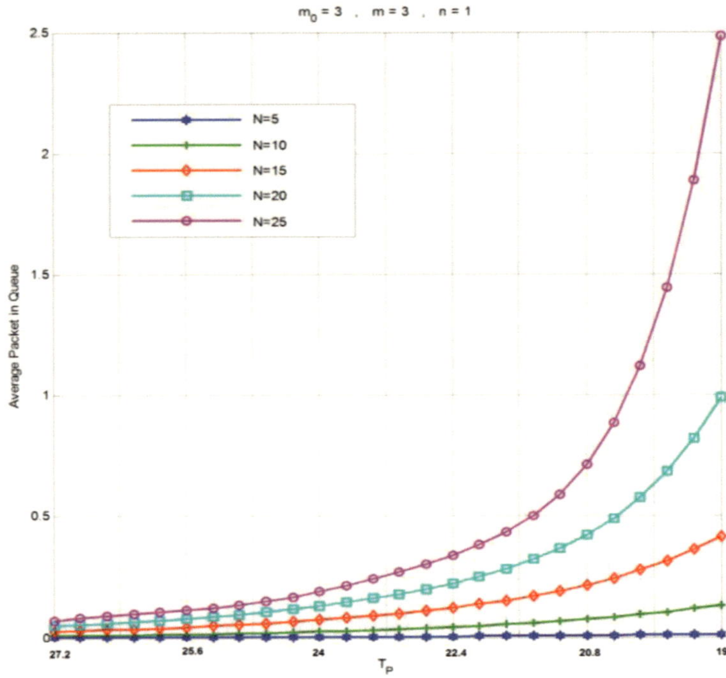

Figure 12.
Average queue length as a function of T_P *and the number of nodes, MAC parameters* $m_o = 3$, $m_b = 8$, $m = 3$, *and* $n = 1$. *Decreasing* T_P *causes an increase in the average queue length, and intensity growth of it depends on the number of nodes in the network.*

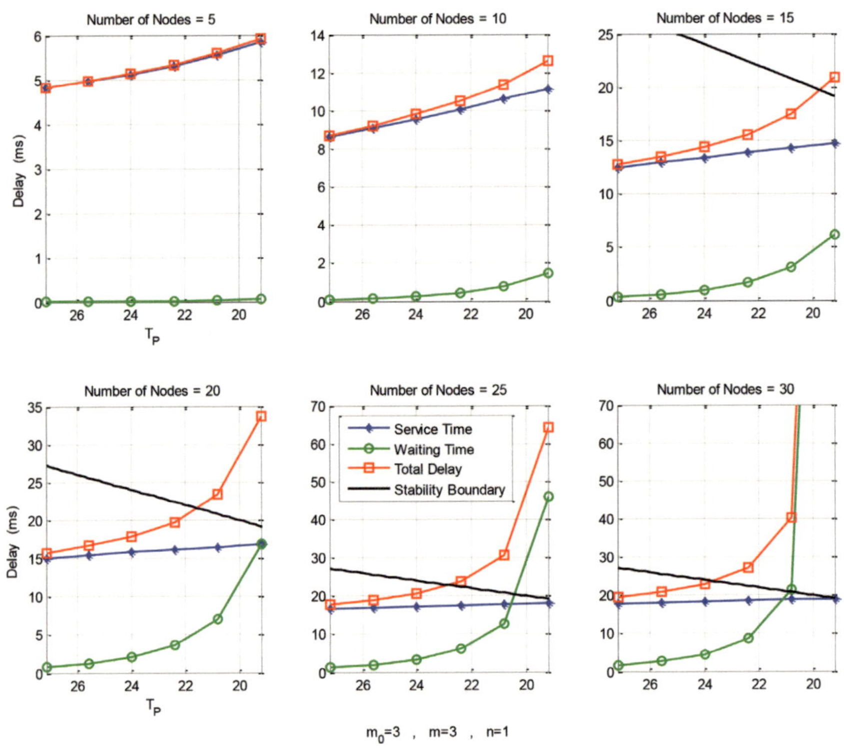

Figure 13.
Service time, waiting time, and total delay against T_P *and N, while* $m_o = 3$, $m_b = 8$, $m = 3$, *and* $n = 1$. *The straight black line determines stability boundary. The delay sensitivity to the time period rests greatly upon N.*

m0 contributes to an increase in average service time, reliability, and maximum value of service time.

The maximum value of the service time, which is a critical factor for the average and peak of the queue length, is specified by PDF's variance. The head area in PDF (the range of values where the PDF is relatively high) has a direct relationship with m_0. It is also notable that the number of probabilities in the head of PDF equals 2^{m_0}. Owing to the uniform distribution of choosing backoff numbers, the slope of the head area is linear.

On the other hand, most of PDF's area is in its head, and due to high reliability, it can be deduced that most of successful transmissions are located in the head area:

$$
\begin{aligned}
E(T_{ServiceTime}) &= \frac{\overline{T}_{successful}P(successful) + \overline{T}_{failure}P(failure)}{P(successful) + P(failure)} \\
\lim_{reliability \to 1} E(T_{ServiceTime}) &= \overline{T}_{successful}
\end{aligned}
\tag{57}
$$

The dotted vertical lines, which correspond to average service time, show the expected value of the corresponding PDF. As this line approaches the head of PDF, the reliability goes up.

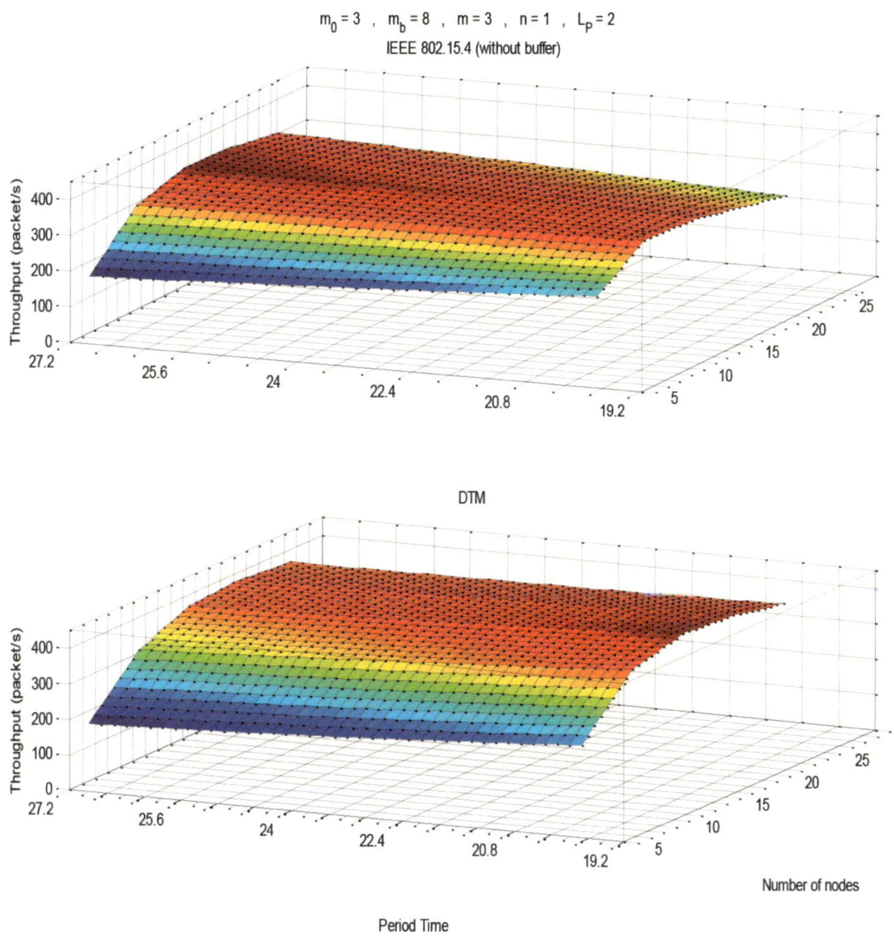

Figure 14.
Throughput as a function of TP and N, while m_o = 3, m_b = 8, m = 3, and n = 1. A comparison is drawn between IEEE802.15.4 and DTM.

On the other hand, average service time increases, while the line moves to the right of the diagram.

As **Figure 11** shows, a reduction in T_P translates into a slight rise in average service time, making the average service time line (dotted red line) far away from PDF's head. As mentioned before, this causes a drop in the reliability. On the other hand, a fall in T_P leads to a slight growth of variance. Change in variance causes fluctuation in the queue average and peak.

As inferred from **Figure 11**, the transition variance from 2.326 to 6.999 makes the queue length and queue peak change from 6.35×10^{-8} and 1 to 2.2451 and 23, respectively.

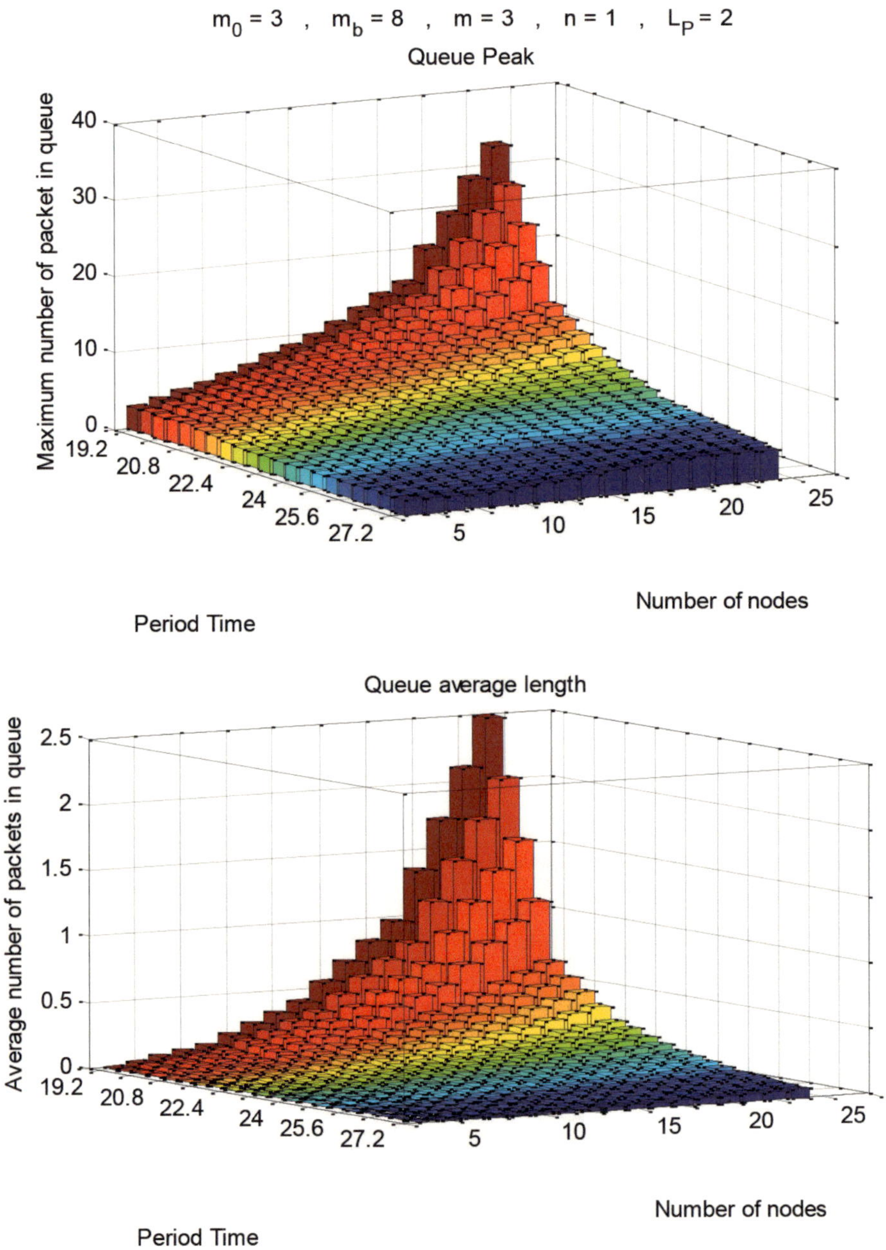

Figure 15.
Queue average and peak against TP and N, while $m_o = 3$, $m_b = 8$, $m = 3$, and $n = 1$.

The more distance between the average service time line and PDF's head we have, the less reliability would occur. As long as the average service time approaches T_P (the dotted red line approaches the dashed black line), the average queue length and peak will exponentially increase. So as to fulfill a stable condition, the dashed line must be on the right side of the dotted line. The less area on the right side of T_P will cause the less queue length average. Hence, T_P must be more than the maximum service time to design a system with zero queue length.

Figure 12 illustrates that the velocity of the increase in the average queue length depends heavily on the number of nodes.

Up to this point, the waiting and transmission blocks were evaluated separately. It is also essential to appraise the effects of both blocks simultaneously. **Figure 13** shows service time, waiting time, and total delay against T_P and N.

A drop in T_P does not affect greatly the total delay in a sparse network (i.e., $N = 5$) but in a dense network does. The straight black line in **Figure 14** determines the stability boundary, so that all network quiescent points upper this line leads the network to an unstable state.

The diagram of throughput against T_P and N is shown in **Figure 14**. To specify the contribution of the queue in throughput, this figure draws a comparison between DTM and IEEE802.15.4 (without buffer). The maximum throughput in DTM (442 packet/s) has considerably increased compared to IEEE802.15.4 (353 packet/s). Furthermore, the way in which throughput rises has changed in DTM. In this simulation, the lower value of throughput corresponds to a network with $N = 5$ and $T_P = 27.2$ ms for IEEE 802.15.4. An increase in N and T_P is responsible for a rise in throughput until a maximum value, in $N = 15$ and $T_P = 27.2$. The reduction in throughput starts following this point. Nevertheless, the throughput in DTM changes in a dissimilar way, so that the maximum throughput is acquired in $N = 15$ and $T_P = 19.2$.

Two factors prove contributing to design a buffer, the average length and peak of the queue. **Figure 15** illustrates the corresponding queue average length and peak of **Figure 14**. Deducing from **Figure 15**, so as to reach the maximum throughput ($N = 15, T_P = 19.2$), a 10 packet length buffer is required in order not to lose any packets.

7. Conclusion

CPSs, developing rapidly and covering eclectic domains, constitute thriving solutions for Smart Grid, the next-generation power grid systems. In this paper, we proposed a novel analytical model based on Markov chain for the MAC sublayer of IEEE802.15.4 standard. This model can provide a precise QoS to applications in which data generation proves periodic, such as AMI in Smart Grid. This is achieved by supplying the model with a MAC-level buffer and the reconsideration of idle mode. The model can provide QoS by reducing the impact of traffic rate fluctuation and the variation of the number of nodes. We incorporated variable idle state lengths so as to makes our study more pragmatic, and then the overall performance in terms of the end-to-end delay and reliability was evaluated. In this paper, the end-to-end delay refers to the interval between when a packet is generated and when a packet service is accomplished, including the time when in the queue as well as transmission time. We observed that the delay distribution of IEEE802.15.4 depends mainly on the MAC parameters and the collision probability.

Furthermore, using the probability density function of transmission time, we designed an optimum network meeting our QoS requirements. We analyzed the impact of MAC parameters and packet generation rate on the shape of the PDFs. In order to make our view more general and feasible, both saturated and unsaturated traffic has been applied, and no limitation is imposed on the queue length.

Besides Monte Carlo simulations, we performed a field test on the protocol by building a WSN with self-designed motes, validating our model. Future work includes investigating the performance of our analytical model with a downlink stream.

Author details

Jafar Rasouli*, Ahmad Motamedi, Mohamad Baseri and Mahshad Parsa
Department of Electrical Engineering, Amirkabir University of Technology, Tehran, Iran

*Address all correspondence to: j.rasooly@yahoo.com

References

[1] Lee EA. Cyber physical systems: Design challenges. In: 2008 11th IEEE International Symposium on Object Oriented Real-Time Distributed Computing (ISORC). 2008. pp. 363-369

[2] U.S. Department of Energy. 2015. Available from: http://www.energy.gov

[3] Saber AY, Venayagamoorthy GK. Plug-in vehicles and renewable energy sources for cost and emission reductions. In: IEEE Transactions on Industrial Electronics. Vol. 58. 2011. pp. 1229-1238

[4] Cecati C, Citro C, Siano P. Combined operations of renewable energy systems and responsive demand in a smart grid. In: IEEE Transactions on Sustainable Energy. Vol. 2. 2011. pp. 468-476

[5] Cecati C, Citro C, Piccolo A, Siano P. Smart operation of wind turbines and diesel generators according to economic criteria. In: IEEE Transactions on Industrial Electronics. Vol. 58. 2011. pp. 4514-4525

[6] Jianhua S, Jiafu W, Hehua Y, Hui S. A survey of cyber-physical systems. In: 2011 International Conference on Wireless Communications and Signal Processing (WCSP). 2011. pp. 1-6

[7] Carreras BA, Lynch VE, Newman DE, Dobson I. Blackout mitigation assessment in power transmission systems. In: Proceedings of the 36th Annual Hawaii International Conference on System Sciences. 2003. p. 10

[8] Pourbeik P, Kundur PS, Taylor CW. The anatomy of a power grid blackout. IEEE Power and Energy Magazine. 2006;**4**:22-29

[9] Erol-Kantarci M, Mouftah HT. Wireless sensor networks for smart grid applications. In: Electronics, Communications and Photonics Conference (SIECPC), 2011 Saudi International. 2011. pp. 1-6

[10] IEEE Std 802.15.4. Available from: http://www.ieee802.org/15/pub/TG4.html

[11] Götz M, Rapp M, Dostert K. Power line channel characteristics and their effect on communication system design. IEEE Communications Magazine. 2004; **42**:78-86

[12] Gungor VC, Sahin D, Kocak T, Ergut S, Buccella C, Cecati C, et al. A survey on smart grid potential applications and communication requirements. In: IEEE Transactions on Industrial Informatics. Vol. 9. 2013. pp. 28-42

[13] Mišić J, Shafi S, Mišić VB. Performance of a beacon enabled IEEE 802.15.4 cluster with downlink and uplink traffic. In: IEEE Transactions on Parallel and Distributed Systems. Vol. 17. 2006. pp. 361-376

[14] Pollin S, Ergen M, Ergen SC, Bougard B, Moerman I, Bahai A, et al. Performance analysis of slotted carrier sense IEEE 802.15.4 medium access layer. In: IEEE Transactions on Wireless Communications. Vol. 7. 2008. pp. 3359-3371

[15] Buratti C. Performance analysis of IEEE 802.15.4 beacon-enabled mode. In: IEEE Transactions on Vehicular Technology. Vol. 59. 2010. pp. 2031-2045

[16] Park P, Di Marco P, Soldati P, Fischione C, Johansson KH. A generalized Markov chain model for effective analysis of slotted IEEE 802.15.4. In: IEEE 6th International Conference on Mobile Ad hoc and Sensor Systems, 2009. MASS'09. 2009. pp. 130-139

[17] Park TR, Kim TH, Choi JY, Choi S, Kwon WH. Throughput and energy consumption analysis of IEEE 802.15.4 slotted CSMA/CA. Electronics Letters. 2005;**41**:1017-1019

[18] Bennett C, Wicker SB. Decreased time delay and security enhancement recommendations for AMI smart meter networks. In: Innovative Smart Grid Technologies (ISGT). 2010. pp. 1-6

[19] Bennett C, Highfill D. Networking AMI smart meters. In: 2008 IEEE Energy 2030 Conference. 2008. pp. 1-8

[20] Güngör VC, Sahin D, Kocak T, Ergüt S, Buccella C, Cecati C, et al. Smart grid technologies: Communication technologies and standards. In: IEEE Transactions on Industrial Informatics. Vol. 7. 2011. pp. 529-539

[21] Fang X, Misra S, Xue G, Yang D. Smart grid—The new and improved power grid: A survey. IEEE Communication Surveys and Tutorials. 2012;**14**:944-980

[22] Hossain E, Han Z, Poor HV. Smart Grid Communications and Networking. Cambridge, UK: Cambridge University Press; 2012

[23] Sood V, Fischer D, Eklund J, Brown T. Developing a communication infrastructure for the smart grid. In: 2009 IEEE Electrical Power & Energy Conference (EPEC). 2009. pp. 1-7

[24] Lu G, Krishnamachari B, Raghavendra CS. Performance evaluation of the IEEE 802.15.4 MAC for low-rate low-power wireless networks. In: IEEE International Conference on Performance, Computing, and Communications. 2004. pp. 701-706

[25] Zheng J, Lee MJ. A Comprehensive Performance Study of IEEE 802.15.4. Los Alamitos: IEEE Press book; 2004

[26] Cao X, Chen J, Sun Y, Shen XS. Maximum throughput of IEEE 802.15.4 enabled wireless sensor networks. In: 2010 IEEE Global Telecommunications Conference (GLOBECOM 2010). 2010. pp. 1-5

[27] Ling X, Cheng Y, Mark JW, Shen X. A renewal theory based analytical model for the contention access period of IEEE 802.15.4 MAC. In: IEEE Transactions on Wireless Communications. Vol. 7. 2008. pp. 2340-2349

[28] Bianchi G. Performance analysis of the IEEE 802.11 distributed coordination function. In: IEEE Journal on Selected Areas in Communications. Vol. 18. 2000. pp. 535-547

[29] Faridi A, Palattella MR, Lozano A, Dohler M, Boggia G, Grieco LA, et al. Comprehensive evaluation of the IEEE 802.15.4 MAC layer performance with retransmissions. In: IEEE Transactions on Vehicular Technology. Vol. 59. 2010. pp. 3917-3932

[30] Zhu J, Tao Z, Lv C. Performance evaluation of IEEE 802.15.4 CSMA/CA scheme adopting a modified LIB model. Wireless Personal Communications. 2012;**65**:25-51

[31] Ramachandran I, Das AK, Roy S. Analysis of the contention access period of IEEE 802.15.4 MAC. In: ACM Transactions on Sensor Networks (TOSN). Vol. 3. 2007. p. 4

[32] Park P, Di Marco P, Fischione C, Johansson KH. Modeling and optimization of the IEEE 802.15.4 protocol for reliable and timely communications. In: IEEE Transactions on Parallel and Distributed Systems. Vol. 24. 2013. pp. 550-564

[33] Al-Anbagi I, Erol-Kantarci M, Mouftah HT. A reliable IEEE 802.15.4 model for cyber physical power grid monitoring systems. In: IEEE

Transactions on Emerging Topics in Computing. Vol. 1. 2013. pp. 258-272

[34] Dorling K, Messier GG, Valentin S, Magierowski S. Minimizing the net present cost of deploying and operating wireless sensor networks. In: IEEE Transactions on Network and Service Management. Vol. 12. 2015. pp. 511-525

[35] Kar P, Misra S. Reliable and Efficient Data Acquisition in Wireless Sensor Networks in the Presence of Transfaulty Nodes. IEEE Transactions on Network and Service Management. Mar. 2016;**13**(1):99-112

[36] Akyildiz IF, Melodia T, Chowdhury KR. A survey on wireless multimedia sensor networks. Computer Networks. 2007;**51**:921-960

[37] Watteyne T, Weiss J, Doherty L, Simon J. Industrial IEEE802.15.4 e networks: Performance and trade-offs. In: 2015 IEEE International Conference on Communications (ICC). 2015. pp. 604-609

[38] Yang D, Xu Y, Gidlund M. Coexistence of IEEE802.15.4 based networks: A survey. In: IECON 2010-36th Annual Conference on IEEE Industrial Electronics Society. 2010. pp. 2107-2113

[39] Kleinrock L. Queueing Systems, Volume I: Theory. New York: Wiley, WileyInterscience; 1975

[40] Bolch G, Greiner S, de Meer H, Trivedi KS. Queueing Networks and Markov Chains: Modeling and Performance Evaluation with Computer Science Applications. John Wiley & Sons; 2006

[41] Lindley DV. The theory of queues with a single server. In: Mathematical Proceedings of the Cambridge Philosophical Society. 1952. pp. 277-289

[42] Spitzer F. The wiener-Hopf equation whose kernel is a probability density. Duke Mathematical Journal. 1957;**24**:327-343

Chapter 6

Microgrid

Nithiyananthan Kannan

Abstract

A microgrid has a group of electrical generation and various types of loads operated as single controllable power system. Microgrid is a best option for configuration of recent model power grids. Microgrids are capable of work in parallel with the existing grid as well as off grid as isolated mode. The microgrid enables the grid connection as either AC grid or DC grid and it provides connections of variable AC and DC sources with loads. Microgrid has modeled such a way that it avoids multiple reverse connections. Power electronic devices such as converters and inverters are ensures safe operation and control of the microgrid. The proper modeling and simulation results ensure the successful implementation of microgrid. The challenges involved in implementation and the modeling of AC/DC and hybrid grid in the tied mode have been discussed. The simulation modeling of the microgrid in MATLAB/SIMULINK platform is explained with neat circuit diagram. This chapter provides the readers complete and comprehensive overview about microgrids and their different modes of operations.

Keywords: microgrid, AC grid, DC grid, renewable energy, MATLAB/SIMULINK

1. Introduction

Energy requirements of the world are increasing day by day since there is a constant increase in the world population. The main source of energy consumed is from oil, gas, and coal. The rate at which fossil fuels are utilized is alarming because the stock of fossil fuels is depleting at a higher rate, and after few years, fossil fuels become out of stock. Due to high demand, the fossil fuel energy cost is very high, and it becomes difficult to extract. Usage of fossil fuels creates problem to the environment and life of the livid hoods in earth. The combustion of oil, gas, and coal produces a pollutant gas which destroys the environment and also contributing to air pollution such as emission of nitrogen, sulfuric, carbon, hydrocarbon, and metal oxides. Oil spills and acid rain are the classical example of this polluted environment. The increase in demand for energy forces to find alliterative fuels has to utilize for green energy production. The best alternative and more suitable for this type of requirement is renewable energy sources. Renewable energy generation source is environment friendly and is available in abundant in nature. This source will not deplete, since the power is extracted from solar, fuel cells, wind, hydro, geothermal, and biomass. From renewable energy sources, energy is derived from direct and indirect way. Now-a-days due to advancement in technology efficiency of renewable energy source, sources improved and cost per unit is reduced. Most of the nations started investing in renewable energy power plants for long-term growth based on their energy growth requirements as shown in **Figure 1**. Since last few years, global investment on renewable energy sources exceeds four trillion billions and in is expected much more in coming years [1–5].

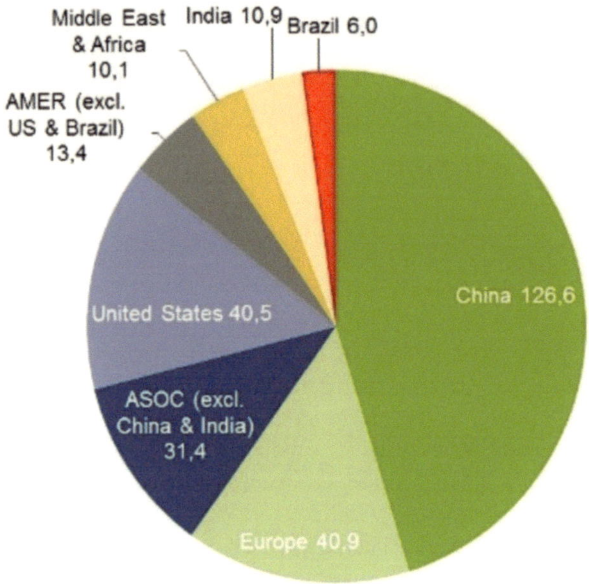

Figure 1.
Global investment in renewable energy in billions.

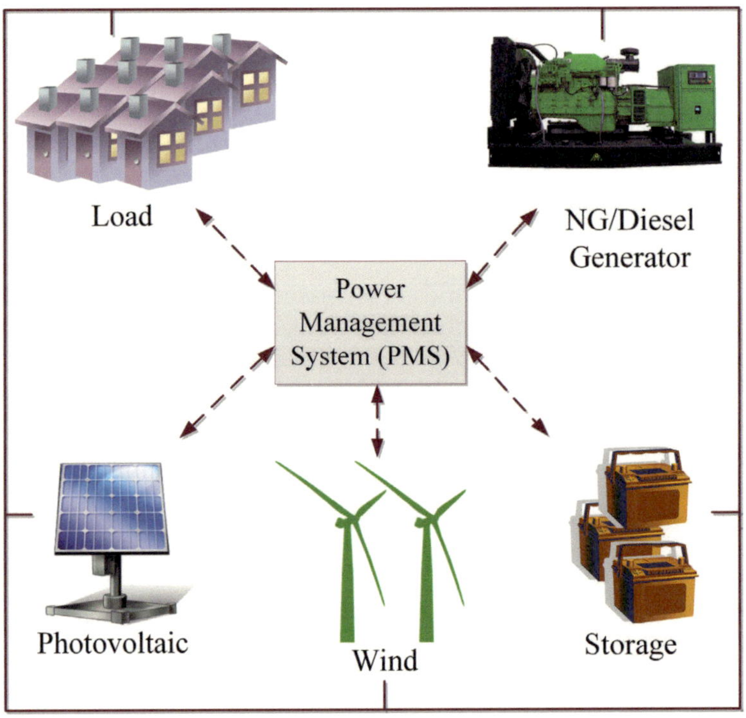

Figure 2.
Microgrid and its components.

With the increase in power demand and raise in the global warming plays a critical role in distributed renewable energy sources. Local generation sources such as wind, biomass, solar, microturbines, and fuel cells are the future renewable sources. Latest power grids have advantages like integration of green energy, scalability, and modularity supported by different renewable and non-renewable energy sources.

Microgrids are the latest evolving grid which effectively extract power and manage in the distributed energy sources. Microgrids are the future grids of the world. The components of microgrids are distributed energy sources such as PV cells, fuel cells, wind turbines, and mini hydro turbines. along with storage devices such as fly wheels, energy capacitors and batteries and diesel generators for peak load management as shown in **Figure 2**.

Microgrids are operated in two modes such as grid connected mode and islanded mode. Microgrids are low voltage grid which can be operated in autonomous state and non-autonomous state. The effective coordination of the components or sources involved in a microgrid will provide energy as minimum cost with high reliability in power supply. Microgrids are the classical example of smart grid applications. The data communications in all nodes of microgrids make the power management much more effective and accurate. Microgrids are low voltage, low power grids ranging from few kW to hundred kW. Based on the size of the load specifications and its applications, microgrids are called as mini grids, house grids, university grids, army grids, and nanogrids [6–13].

2. Technical challenges in microgrid

The challenges related with the control and operations of microgrids are very high. Microgrids are required to ensure the reliable operations even at fault conditions, power system stability during disturbances, and power quality in the island mode. The grid connected microgrid is needed to be maintained synchronism at any situations. Microgrids need to have advanced control strategies for microgrid inverters to maintain correct frequency and voltage to ensure stable operations for the power system connected with dynamically variable load. The growth of microgrids and its challenges has gained attention of researchers and various government and private organizations to play a vital role in finding out the solutions of effective implementation of microgrid. One of the major problems in microgrid is protection system in main grid and microgrid faults. When there is any fault in main grid protection, microgrid needs to be isolated immediately, and similarly if there is any fault in microgrid, the protection system should isolate the smallest part of the microgrid to clear the fault. The selectivity and sensitivity plays a vital role in low voltage microgrid power management. False tripping, unnecessary tripping, delayed tripping, and undetected faults are the major challenges in design of protection system in microgrids. The number of installation of distributed energy resources and availability of short-circuit current in the islanded mode of microgrid are the major issues while considering the protection of microgrids. Compared to main grid, the short-circuit current will drop down drastically and will give problem in the protection of microgrid. There is a need for designing a proper protection system for distributed energy resources with short-circuit calculations and placement of over current relays, reverse power relays, and directional over current relays. In real time, the operating conditions of microgrid are variable because of intermittent distributed sources and dynamic electrical load demand. This leads to change in network topology frequently to aim to minimize loss, economic load dispatch, and proper unit commitment with satisfying all the constraints. The directions and magnitudes of short circuit will vary because of these situations. The different sizing of equipment in various components of microgrid creates often a loss of relay coordination, and generic over current protection will become ineffective in protection of the microgrid from faults. The invention of new methodology in protection mechanism is very much needed to set different parameters for over current protection, parameters of relays, deal with low short circuit current, and interfacing of power electronic devices with the microsources [14].

3. Microgrid benefits

The evolution of microgrid is able to provide solution to problem of integrating huge amount of micro-energy sources without affecting the main power supply from the power utility provider. With the intelligent controllers, microgrid works effectively with the existing power system distribution services to match the variable load demand. Some of the times, microgrid can act as an auxiliary for the main grid to support the distribution and utilization. With the support of protection systems is able to disconnect from the disturbance if there is any fault in main grid, and microgrid is able to operate in a standalone mode which improves the reliability and quality of the power to the consumer. Microgrid is a decentralized grid with improved efficiency, located near to the load, and capable of extracting power from distributed renewable energy sources. Microgrid special features are highlighted in **Figure 3**.

Consumers able to get advantages from microgrid because it is developed and utilized for the local needs in terms of temperature and energy consumption, able to provide uninterruptible power, highly reliable, reduced losses, and improved voltage profile. Apart from the power generation and distribution microgrid involves in storage, control of load profile and extracting energy from heat. The success behind microgrid is capability of the grid to have smooth mode change from island to grid and vice versa.

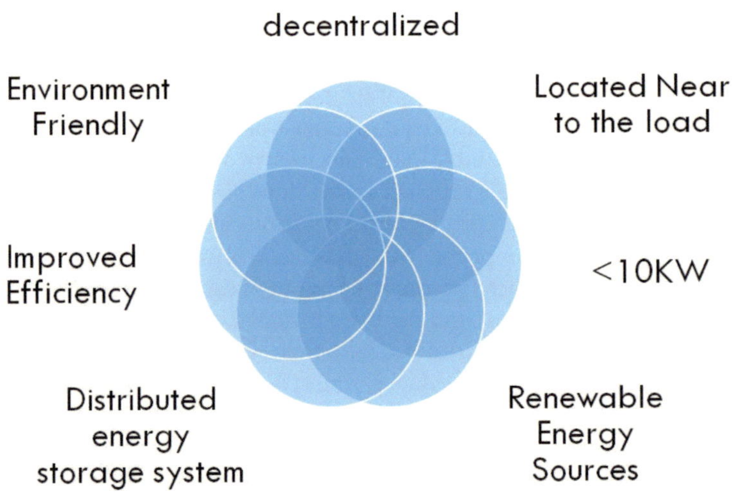

Figure 3.
Microgrid features.

4. Renewable energy sources

Back bone of the microgrid power supply is renewable energy sources. Most commonly used renewable energy sources are fuel cells, microturbines, solar, and wind. Fuel cells can produce power through electrochemical actions among oxygen and hydrogen. This reaction is highly effective and gives bi product as water and temperature. Due do this environment friendly bi products, this technology becomes popular compared to conventional diesel engines which are more expensive and pollute environment by releasing flue gases. Even though fuel cells are discovered in eighteenth century, it has been in use in late nineteenth century. Because of usage in commercial and domestic applications, fuels cells power technology become popular. Fuel cells have more advantages such as power quality, very

Figure 4.
Bloom box servers.

Figure 5.
Wind turbines.

high efficient, environment friendly, and modularity. Recent times in microgrid applications, fuel cells are very effective in replacing IC engine usage which makes microgrid environment free. Further fuel cell technology usage will reduce greenhouse gases and improved efficiency of the microgrid. Advancement in fuel cell technology leads to bloom box energy servers which convert fuel into electricity through electrochemical approach. Bloom box servers are available as building box as clusters ratings from hundred kilowatts to megawatts. Bloom box is highly efficient in power production, modular, and scalable (**Figure 4**).

Wind energy is most promising sources of renewable energy source, which has more potential in future. Wind energy power plants are sustainable energy producers. Wind turbines play a vital role in supplying power along with fuel cells and solar cells to microgrids. Wind turbines are bundled into components such as rotor, generator, turbine blades, and driver or a coupling device as shown in **Figure 5**. As high speed rotates the blades, the air exerts aero dynamic forces will make the blades to turn the wind rotor. In turn rotor rotates the generator through gear box. Wind generators are mostly induction generators or permanent magnet generators which supplies power to the microgrid.

Solar power plant generation systems are conversion of sunlight into electric power either directly using PV cells or indirectly using concentrated solar power plants.

Figure 6.
Solar power plants.

Photovoltaic generation is systems, which convert the sunlight directly to electricity as shown in **Figure 6**. PV cell technology is very well established and used extensively in microgrid. The DC output of the connected PV cells is given as input to the inverter and AC power supply given to the load through inverter. The most common renewable energy source in the world is solar energy.

All microgrids are more effective in extracting local energy source is only from solar. The main advantages of these solar power plants are very less time required to design and install the power plant. These plants are highly modular, and it is good alternative for peak load demand. Since solar structures are static and no moving parts which gives no noise power plant status to solar power plants. Solar power plants are portable and mobile because of light weight. Since there is no moving part, PV generation systems are having longer life time which makes one of the attracting renewable energy sources connected with microgrid.

Microturbines and microhydro turbines are gas and hydro electric generators ranging in size from 25 to 500 KW connected with microgrid. These microturbines are very much useful in dealing with peak load demands for the microgrids. Biomass is also another renewable source contributing energy supply to microgrids located in remote villages or urban waste sources. The source materials are scrap lumper, forest debris, certain crops or manure, etc. Geo thermal power plants are recent renewable energy sources generating power from geo thermal energy. Geo thermal energy is a thermal energy stored in the earth is utilized. In future, microgrids are expected to add mini geo thermal power plants in its pool of distributed energy resources. In future, the other renewable energy power supply from tides of the sea and hot hydrogen fusion will be a part of microgrid energy sources.

5. AC/DC and hydro microgrid

Microgrids are the most prominent grids to extract various types of distributed renewable energy sources with the distribution grid. The distribution utility side grids are AC grids, since today electrical loads are connected with power electronic devices, and since most of the renewable energy sources are DC power supply, now the DC microgrids concepts are gaining more attraction. The most common grids are AC microgrids which uses standard protection technique's to manage disturbances. But the distributed generation sources for microgrid comprising of solar cell arrays and fuel cells are DC power which needs to convert to AC power through power electronic device. This DC power source in the microgrid is connected to the power supply through inverter to the utility grid. The AC power on the utility side is converted back to DC for some of the latest electrical loads such as UPS batteries, DC lighting loads, DC motor drives, and hybrid electric vehicles. Due to multiple conversions, AC microgrid becomes inefficient due to more power losses due to multiple conversions. Apart from that maintain synchronism, stability issues and reactive power requirements are the challenges faced by AC microgrids. Due to this reason, DC microgrids are emerging as alternate grids. But the diesel generator, small microturbines connected synchronous generator, and wind energy generators are required to connect with AC/DC converter to distribution side grid. Due to multiple conversions, DC microgrid also becomes inefficient due to more power losses due to multiple conversions. The reason for this problem is now-a-days electrical load becomes mix of both AC and DC power. It leads AC/DC grids to less efficient operation. The best solution to avoid this multiple conversion energy losses is hybrid AC/DC microgrid. The hybrid AC/DC microgrid is formed with an objective to minimize the conversion losses to make the microgrid more efficient as shown in **Figure 7**. But the implementation of hybrid of AC/DC grid required several technical challenges.

Hybrid microgrid comprises both AC and DC sources as shown in **Figure 7**. The respective AC and DC sources are connected to the corresponding AC and DC networks. The AC bus and DC bus are linked together with three phase converters and the transformers. The AC bus of hybrid grid is tied with distribution grid with transformer and circuit breaker. The successful operations of hybrid microgrid are based on power conditioning converters. The converters are classified based on their input and output power supply. The converters used in hybrid microgrids are rectifier, inverter, buck/boost converter, and transformers. Among these converters, inverters and boost converters need more attention than conventional converters in the microgrid. Because the boost converter is very much required to interface low voltage PV arrays into the microgrid and then converted into AC power through inverter to feed the loads. **Figure 8** shows how the converters are theoretically

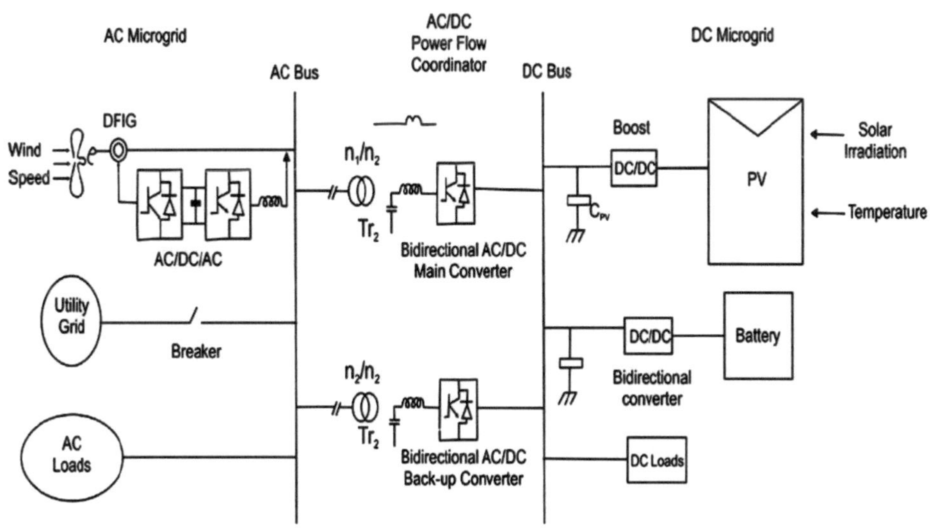

Figure 7.
Hybrid microgrid.

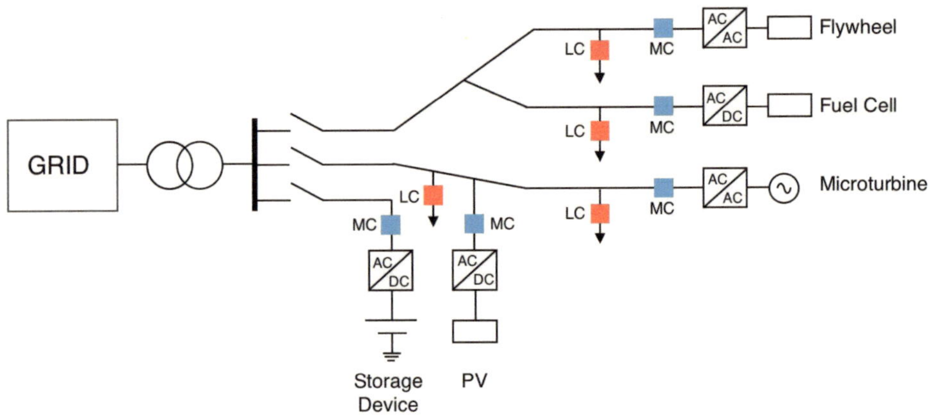

Figure 8.
Implementation of converters in a typical microgrid.

implemented; in this Figure, LC stands for load controller and MC for microsource controller. Arrow indicates the electrical load.

5.1 Inverter

Inverter is a power electronic device, which converts DC to AC. The design of the inverter fixes the values of the input voltage, output voltage, frequency, and power output. An inverter is classified into single phase and three phase inverter. The single phase full bridge inverter consists of four switches, and it creates an AC output voltage by switching on and off as shown in **Figure 9**.

Three single three-phase inverters form the three phase inverter. Three-phase inverter has six switches each switch operated at 60 degree point of the fundamental output waveform. Based on the switching frequency and switching pattern, the output voltage, frequency, and power output have been controlled. **Figure 10** shows the general standard equivalent circuit of a three-phase inverter interfaced with solar panels output to microgrid.

5.2 Boost converter

A boost converter is used as DC to DC power converter to increase the voltage and in turn it will step down the current. In microgrid, one of the major sources is solar energy with DC power supply. The PV panel output voltage is not suitable to feed the loads straightaway. The boost converter is used to step-up the voltage to a required level. **Figure 11** shows the equivalent circuit of boost converter in two modes. One mode switch is closed and other mode switch is open.

Based on the operation of switches, the boost converter operations are divided into two modes. When the switch is closed, the energy is stored in inductor as magnetic energy. Since the diode is switched off, the capacitance is blocked from the power supply. When the switch is open, the inductance is then series with the source which enhances the voltage as boost voltage. The boost voltage is depending on the value of inductor and capacitor.

5.3 Operating modes of the microgrid

Microgrids are operated in two modes, one is grid connected and another one is islanded mode. The first one is the classical scheme, which is the most

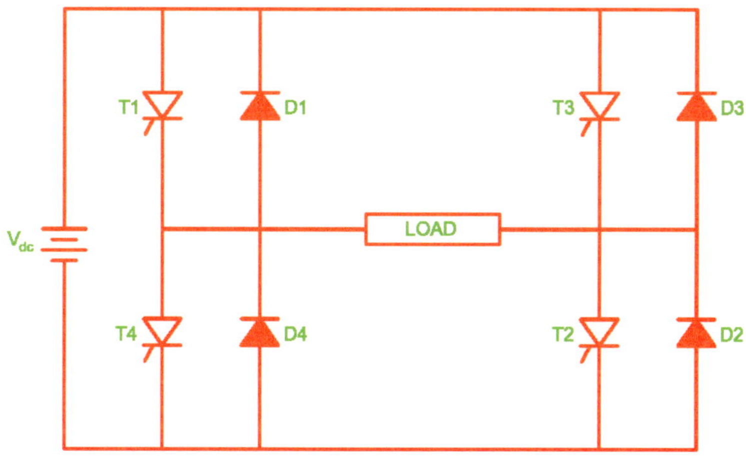

Figure 9.
Equivalent circuit of a single-phase full bridge inverter.

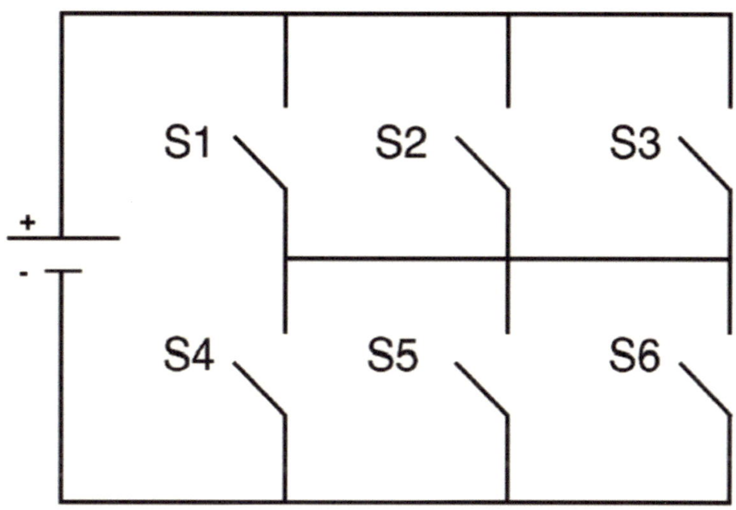

Figure 10.
Equivalent circuit of a three-phase inverter.

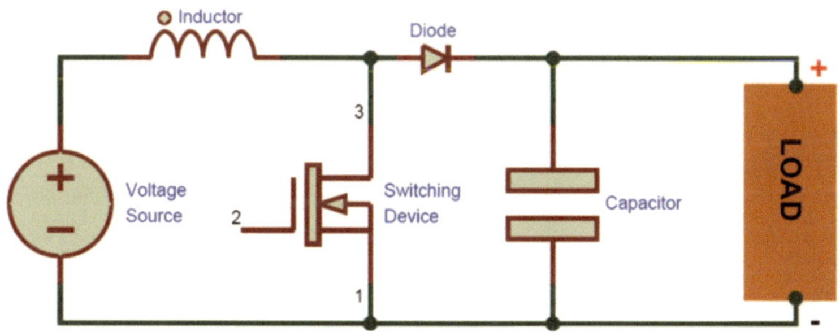

Figure 11.
Equivalent circuit of a boost converter.

common mode in use. Microgrids are designed to operate in both modes. Microgrids are designed such a way to get maximum power from renewable energy sources with wind, solar, and microturbines and able to fill its remaining required power from the grid. This mode of operation ensures maximum possible power to extract from local sources ensures the cost of energy cheapest. In the second mode, when the network gets isolated to faults, emergency conditions or natural calamities, the microgrid designed to operate in isolated mode. In the isolated mode, the entire power is supplied from locally available power sources of the microgrid.

6. Microgrid modeling and simulations

The MATLAB/SIMULINK-based modeling is used for designing a microgrid. Based on the individual microgrid specifications, model has been formed and implemented in MATLAB/SIMULINK environment. A hybrid microgrid sample system has been implemented in the MATLAB/SIMULINK environment as shown in **Figure 12**. In this model, microgrid is carried out for the grid connected mode. Along with this microgrid model, the performance of the wind generator and solar panel output is also analyzed. While considering the performance analysis of the simulation model, the solar irradiation, temperature of the panels, wind speed, and wind direction also possible to consider getting the results accurately.

The simulated results are analyzed in the runtime environment of MATLAB. Normally for the performance analysis, the load flow of the microgrid has been carried out at one-day simulation results. By running the simulation model in the MATLAB/SIMULINK environment at different day environment and performance of the microgrid has been analyzed. The stage wise output characteristics such as PV panel output, battery, and converter gate signals of the simulation has been represented as **Figures 13–16**, respectively. The output characteristics of AC load voltage and current are represented by **Figures 17 and 18**, and also the **Figures 19 and 20** indicate the converter side AC voltage and current. The simulations need to perform many times repeatedly under various simulated situations, and the final pool proof design needs to be finalized. All the figures output indicates the performance under one similar situation which is useful for the readers to understand the operations of microgrid.

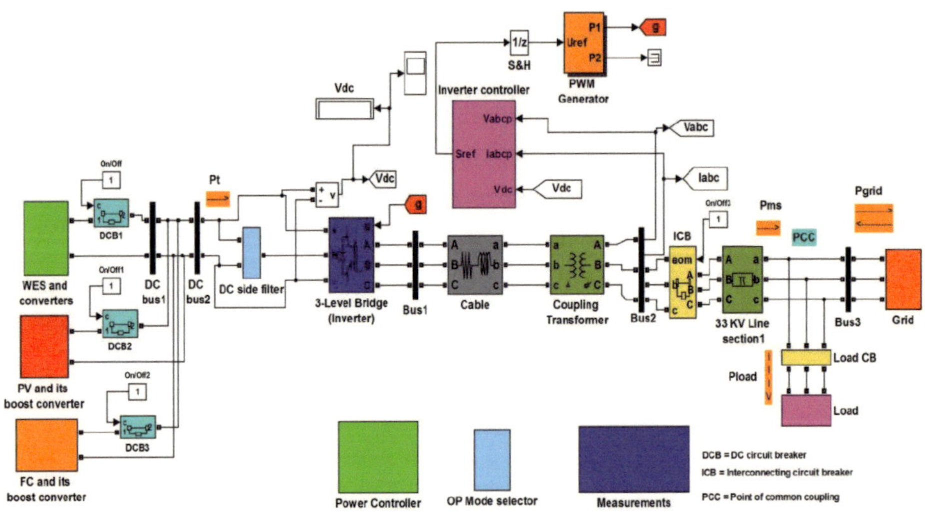

Figure 12.
MATLAB/SIMULINK model for microgrid.

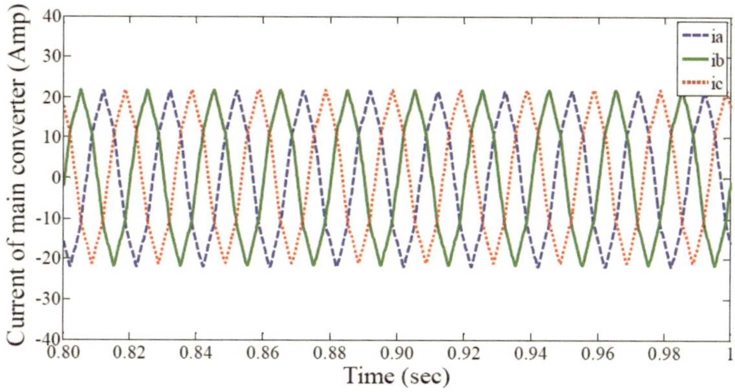

Figure 13.
Output voltage of PV array.

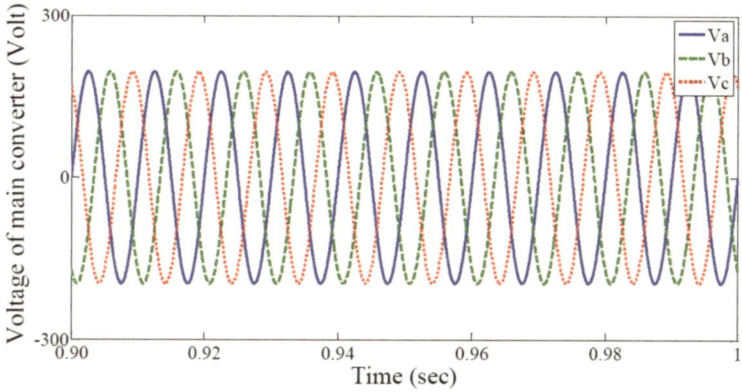

Figure 14.
Output current of PV array.

Figure 15.
Generated PWM signals for the boost converter.

In this hybrid microgrid during one day simulation, it is needed to phasor power gui instead of discrete power gui in MATLAB environment. The individual components are specified with their values in the blocks. The battery block in the MATLAB/SIMULINK model is of 300 Ah capacity and PV panels arrays are modeled by a current source and load current is specified based on the input power. The power management and voltage profile are done for the hybrid microgrid controlled

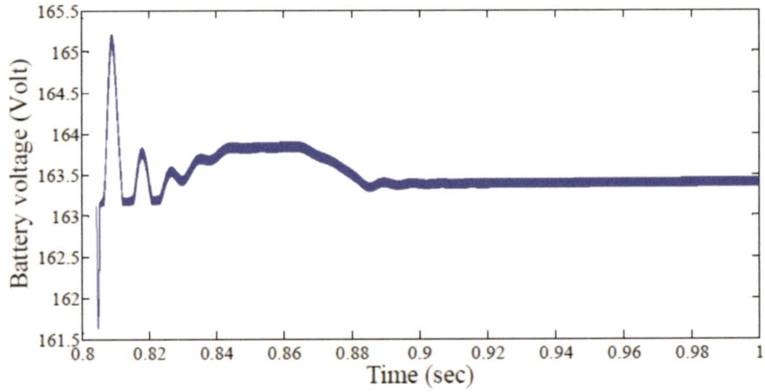

Figure 16.
Voltage of battery.

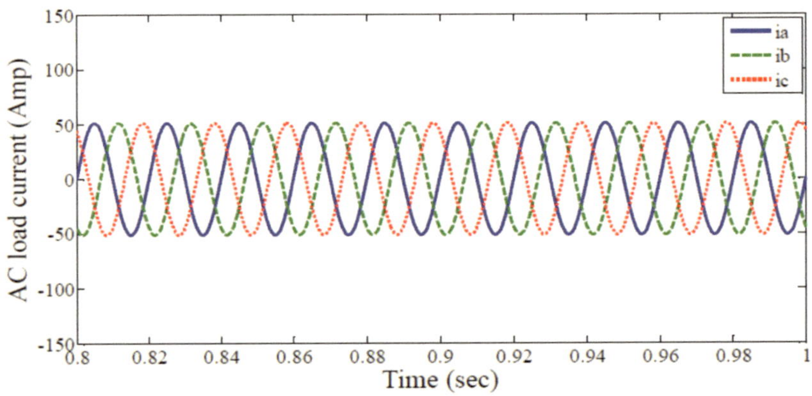

Figure 17.
Microgrid AC bus load current.

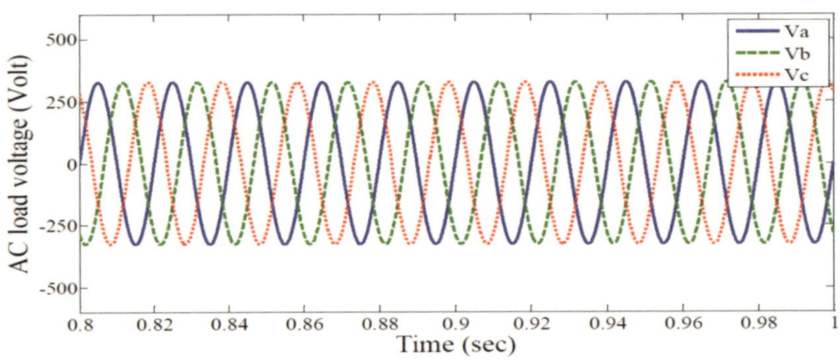

Figure 18.
Microgrid AC bus load voltage.

by PI controller. Once the performance analysis of the simulation model is satisfied, the microgrid can be implemented in real time. Many microgrids are implemented across the world and for example the real time implemented the microgrid as shown in **Figure 21** [15].

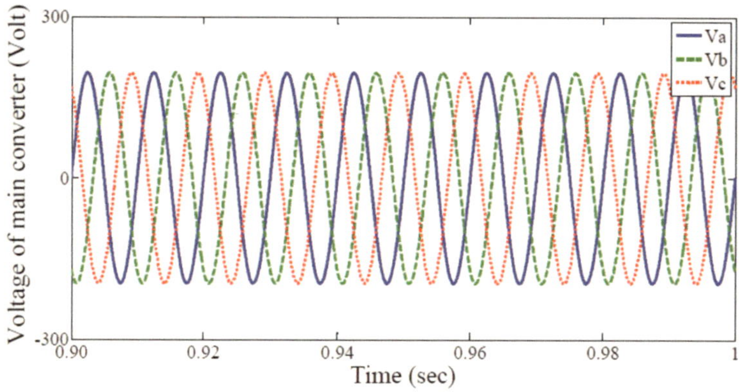

Figure 19.
AC side voltage of the main converter.

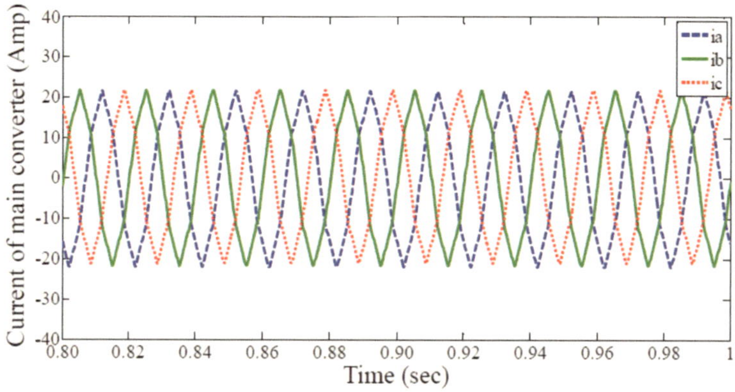

Figure 20.
AC side current of the main converter.

Figure 21.
Microgrid in real time [15].

7. Conclusion

The implementation of microgrids reduces the energy cost and increases reliability. Microgrids are excellent in extracting energy from renewable energy. Microgrid is a controlled system which offers both power and heat. The heat collected by solar thermal collectors is used to heat the air as well as water, which reduces the burden of microgrid energy supply which is becoming popular in residential and commercial loads. Distributed power source is becoming more attractive because the green energy, pollution free, modular, and availability are in plenty, and it can supply power to the existing transmission and distribution systems without any alterations. Microgrids are excellent system for extracting power from distributed generation. In otherwise, microgrids are operating as a locally controllable cell of power system network. Microgrid operations are done by grid connected and islanded mode. In grid connected mode each individual, the micro-sources able to act as current source with maximum power transfer capability to the grid. Any fault or emergency situation microgrid will operate under islanded mode automatically which ensures the reliability of the power supply. The implemented simulation model for the hydro microgrid is explained in a MATLAB/SIMULINK environment. Although the hybrid grid model replaces the AC and DC grid, there is a technical challenge for the implementation of the hybrid grid model based on the AC dominated utility electrical setup. The peer-to-peer and plug-and-play model are used in the microgrids, which ensures that any component loss will also make the microgrid operational and it can continue the power supply. The data communications and control mechanism implementations make the hybrid microgrid as smart grid. The microgrid is acting as application of smart grid which is the future grid of the world.

Author details

Nithiyananthan Kannan
Department of Electrical Engineering, Faculty of Engineering, King Abdulaziz University, Rabigh, Kingdom of Saudi Arabia

*Address all correspondence to: nithiieee@yahoo.co.in; nmajaknap@kau.edu.sa

References

[1] Chen H, Leng H, Tang H, Zhu J, Gong H, Zhong H. Research on model management method for micro-grid. In: IEEE 2nd Information Technology, Networking, Electronic and Automation Control Conference (ITNEC), Chengdu; 2017. pp. 163-166

[2] Chen Z, Wang K, Li Z, Zheng T. A review on control strategies of AC/DC micro grid. In: IEEE International Conference on Environment and Electrical Engineering and 2017 IEEE Industrial and Commercial Power Systems Europe (EEEIC/I&CPS Europe), Milan; 2017. pp. 1-6

[3] Zheng H, Ma H, Ma K, Bo Z. Modeling and analysis of the AC/DC hybrid micro-grid with bidirectional power flow controller. In: China International Electrical and Energy Conference (CIEEC), Beijing; 2017. pp. 280-284

[4] Dongmei Z, Nan Z, Yanhua L. Micro-grid connected/islanding operation based on wind and PV hybrid power system. In: IEEE PES Innovative Smart Grid Technologies, Tianjin; 2012. pp. 1-6

[5] Li Z, Li Y, Fu G, Li B. Directional protection based on fault component energy function in micro-grid. In: IEEE PES Innovative Smart Grid Technologies, Tianjin; 2012. pp. 1-5

[6] Liu Z et al. Typical island micro-grid operation analysis. In: China International Conference on Electricity Distribution (CICED), Xi'an; 2016. pp. 1-4

[7] Jahanghiri H, Rahimi S, Shaker A, Ajami A. A high conversion non-isolated bidirectional DC-DC converter with low stress for micro-grid applications. In: 10th International Power Electronics, Drive Systems and Technologies Conference (PEDSTC), Shiraz, Iran; 2019. pp. 775-780

[8] Zhaoyun Z et al. Application of micro-grid control system in smart park. The Journal of Engineering. 2019;**2019**(16):3116-3119

[9] Deshpande K, Karnataki K, Darshana P, Deshpande MH, Shankar G. Smart renewable energy micro grid for Indian scenarios. In: International Conference on Advanced Computing and Communications (ADCOM), Chennai; 2015. pp. 22-26

[10] E S N Raju P, Jain T. Hybrid AC/DC micro grid: An overview. In: Fifth International Conference on Power and Energy Systems, Kathmandu, Nepal; 28-30 October, 2013

[11] Mani P, Nithiyananthan K, Nair P. Energy saving hybrid solar lighting system model for small houses. World Applied Sciences Journal. 2015;**33**(3):460-465

[12] Jacob D, Nithiyananthan K. Smart and micro grid model for renewable energy based power system. International Journal for Engineering Modelling. 2009;**22**(1-4):89-94

[13] Thomas S, Nithiyananthan K. A novel method to implement MPPT algorithms for PV panels on a MATLAB/SIMULINK environment, Journal of Advanced Research in Dynamical and Control Systems. 2018;**10**(4):31-40

[14] Priyadarshanee L. Modeling and control of hybrid AC/DC micro grid [thesis]. Rourkela, India: National Institute of Technology; 2012

[15] https://new.abb.com/news/detail/4482/abb-ability-supports-indias-clean-energy-future

Chapter 7

Solid-State Transformer for Energy Efficiency Enhancement

Fernando Vaca-Urbano, Manuel S. Alvarez-Alvarado, Angel A. Recalde and Félix Moncayo-Rea

Abstract

The rapid evolution of power electronic solutions in all around the globe brings a common problem, which is the adoption of nonlinear loads. This fact carries out a strong impact over the quality of power systems and consequently on energy efficiency, since nonlinear loads act as sources of harmonic currents that flow to other loads or even sources, causing non-optimal performance in their operation. Nowadays, conventional transformers are limited to just manage (increase or decrease) voltage level, but they are not able to deal with power quality events, such as harmonics, sag, swell, among others. Hence, there is a need to incorporate a versatile smart device to deal with the challenges previously described for a smart grid environment. This chapter introduces a solid-state transformer (SST) with topology of multilevel cascade H bridge converter as a solution. SST is an emerging technology that has the advantages of low volume, low weight, fault isolation, and other management features. Within its fundamental operation, this chapter presents a detailed description of a SST system comprising communication and control, highlighting their main advantages in comparison with conventional transformer such as mitigation of waveform harmonic distortion, allowance of integration of distributed generation, and bi-directional power flow.

Keywords: energy efficiency, nonlinear loads, power quality, smart grids, solid-state transformer

1. Introduction

Electricity is the fundamental enabler of human development. It permits technological advancements that are reflected in constant growth, while expanding its usage as demands increase. In order to produce electricity, several forms of energy with relatively abundant resources have been harnessed, such as hydroelectric, fossil fuels, and nuclear. However, major economic and sustainability factors throughout history have driven the energy consumption balance toward the exploitation of renewable energies. In fact, as these alternative types of primary energy are available at a variable rate defined by uncontrolled weather, its integration to the electrical network must accomplish a high level of control complexity in order to maximize generation without compromising grid safety.

Traditionally, power systems have had functional topologies that have served convenient routes from bulky generation power plants to load consumption centers. Moreover, distribution systems have been mostly designed radially although there are also possibilities to transfer circuits to other feeders in case of unscheduled disconnections. Despite the fact that the electrical network was conceived to transport energy vertically from generators to load, it has faced a major challenge to cope with the advent of renewable energies: the bi-directionality of the energy flow [1]. This characteristic aims to provide renewables sources to be distributed across the network in different sizes and at different locations, at the cost of increasing the number of interconnections in the distribution system, introducing new devices, and redesigning existing implementation practices. This paradigm has been fundamental to envisage the concept of smart grid [2], not only because of the energy shift but also the added intelligence the system must have to control such distributed scenario [3, 4]. Additionally, smart grids can contribute to grid survival in the case of natural disasters and large power plant blackouts. Thus, sustainability and safety are concepts that must fit in the smart grid landscape.

Although the distributed energy across the power system has technical advantages so far, much has to be done in order to make it stable and comply with operational and quality standards. There have been several approaches to study reliability improvements [5–8], stability performance [9], communication technologies [10], and several other organizational transformations [11]. As an illustration, under the operational requirements for power system protection, fault-tolerant systems must discriminate the type of failure event based not only on its own measurements but also on its proximity. Thus, integrated communications systems are of uttermost importance in this case [12]. On the other hand, power quality issues must be compensated because other types of phenomena rising from the utilization of new switching technologies based on power electronics will emerge. Hence, maintaining voltage, frequency, and signal cleanliness even during rare extremely low probability events will become a must in new electrical energy devices. If all these conditions are met, the network operator can ensure the stability of the more complex power system. The future smart grid is an intelligent grid with higher levels of reliability and efficiency [13]. Some of the challenges that the smart grid must manage are detailed in **Table 1**.

The newly demanded performance and functionality mentioned above cannot be obtained with current low frequency power transformers in the grid. These devices transform transmission medium-voltage electrical energy to consumable low-voltage electrical energy at 50/60 Hz frequency. Although it has proven to be highly reliable since power electrification days, it is not designed to handle distributed energy DC production and bi-directional power flow and does not have the capability to handle more complex control other than connection, disconnection, or voltage magnitude control (e.g., tap changers). Nevertheless, the distribution grid had an impulse of intelligence with the deployment of reclosing devices, thus adding more components to the existing infrastructure. This feature has been improved in the last three decades in order to adapt to more stringent conditions.

Challenges	Application
Safety	Wide area monitoring (e.g., fault location)
Greener resources	Integration of dispersed renewable generation and bi-directional customer (utility relation)
Power operation	Improved demand control
	Automated power system operation
	Energy quality improvement

Table 1.
Some challenges for smart grid.

But the introduction of local generation, power electronic devices, higher power requirements, and energy storage proved the current grid to be unable to handle all the operational challenges. Therefore, a new highly controllable modular device is needed to comply with the added complexity [14] of the network while maintaining quality standards. The solid-state transformer (SST) has shown to be flexible enough to accommodate several complex functionalities at different voltage levels with the advantage to be lighter and more efficient than the conventional power transformer and its recloser counterpart.

In fact, the SST provides the following features: availability of low-voltage DC link, power factor correction, VAR compensation, active filtering, disturbance isolation, and smart protection. The DC link allows the direct injection of distributed renewable energy into the grid. On the other hand, its other features add improved compensation and stability [15] for active and reactive power flow within a single smart device.

The SST concept is not only promising for the smart grid but also for other engineering applications. There have been successful attempts to introduce SST for traction process such as railway transportation, remotely operated vehicles (e.g., submarine applications for deep water exploration), and ship propulsion. Hence, there are a full spectrum of possibilities in which SST has shown to be a feasible alternative, for that reason it is sometimes known as the future "energy router."

In order to illustrate the aforementioned capabilities of the SST, this chapter provides an insight into the operation of an SST. Nonlinear loads are revisited as an important part of the SST demand. Then, a mathematical model of an SST is detailed, and its performance under typical power system conditions and disturbances are analyzed. Additionally, a communication feature is also described, such that SST could not only be remotely operated but also take coordinated decisions to optimize power system operation and performance. Therefore, in the next sections, the advantages of SST are studied, demonstrating its feasibility for sustainable smart grid applications.

2. Mathematical model

In order to study the benefits of SST in the smart grid, a system model that could represent its electrical properties is needed. Therefore, in this section the mathematical model of nonlinear loads, SST, and some common power system disturbances are described. Later, these models will be used to analyze power system disturbances and highlight the advantages SST has on the network operation.

2.1 Nonlinear loads

Concerning the definition of a nonlinear load, it is necessary to specify linearity. Linearity is a characteristic used to describe linear loads, and it corresponds to a property in which loads exclusively produce fundamental sinusoidal current if supplied by a sinusoidal voltage source at fundamental frequency [16]. In contrast, nonlinear loads provide distorted current waveforms, thus injecting harmonic components in the system [17]. Load harmonics higher than fundamental frequency are commonly represented with a resistance-inductance-capacitance (also known as RLC) circuit in parallel with a current source, as shown in **Figure 1**.

Nonlinear loads act as sources of harmonic currents whose frequencies are multiple of the fundamental frequency. Harmonics circulated from the load to the source and, depending on the topology of the network, harmonic current can

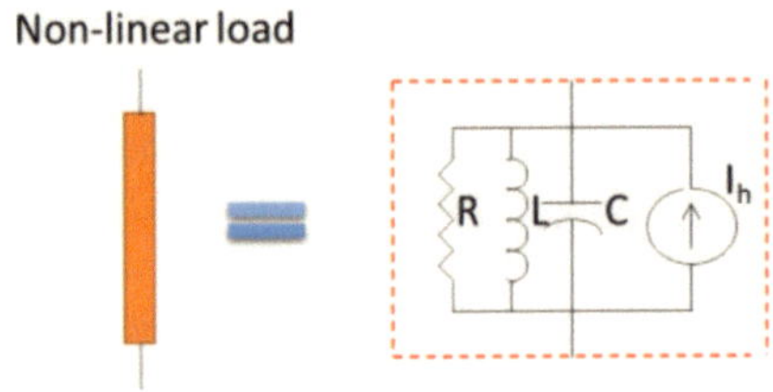

Figure 1.
Equivalent circuit of a nonlinear load.

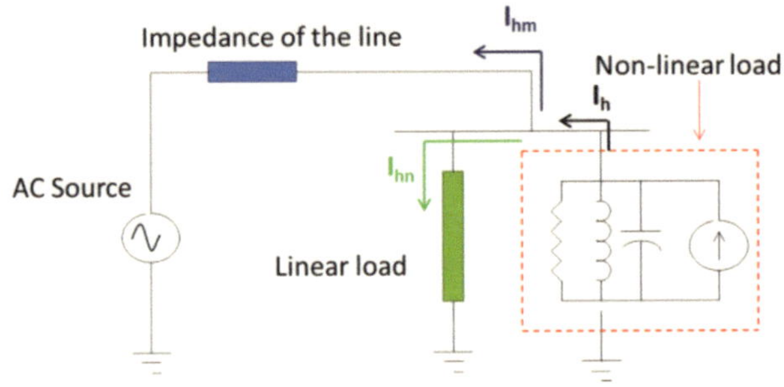

Figure 2.
The effect of harmonics.

spread to other loads. These distorted current components may cause voltage spikes and terrible damage to nearby equipment. Note these phenomena in **Figure 2**.

The fundamental current waveform as a function of time t can be represented as in (1), where the term I_1 represents the current fundamental peak amplitude, ω_0 is the fundamental angular frequency, and θ_1 is the phase angle.

$$i_1(t) = I_1 \cos(\omega_0 t + \theta_1) \tag{1}$$

For simplicity, it is common to represent a sinusoidal function in its phasor form, where it is written as a complex number with amplitude and phase. The phasor amplitude is obtained from the root mean square value of the fundamental sinusoid function amplitude, that is, $I_{1rms} = I_1/\sqrt{2}$ [18]. Applying this criterion to (1), its phasor representation is as follows:

$$\widetilde{I_1} = I_{1rms} \measuredangle \theta_1 \tag{2}$$

In the presence of harmonics, waves are distorted and become a function of the total number of harmonics NH. Considering this fact and using Fourier series, the current has the form as presented in (3) [19].

$$i(t) = \sum_{h=1}^{NH} I_h \cos(h\omega_0 t + \theta_h) = I_1 \cos(\omega_0 t + \theta_1) + \sum_{h=2}^{NH} I_h \cos(h\omega_0 t + \theta_h) \tag{3}$$

Notice that the left-hand side term of the sum is the fundamental frequency sinusoid, which has exactly the form as presented in (1), while the right-hand side term is the harmonic currenti_h, i.e., the distortion wave. Then, (3) can be rewritten as:

$$i(t) = i(t)' + i_h(t) \tag{4}$$

By expressing (4) in its phasor form, the result is as given in (5).

$$\widetilde{I} = I_{1rms}\measuredangle\theta_1 + I_H\measuredangle\theta_h = I_{rms}\measuredangle\theta_I \tag{5}$$

where $I_H = \sqrt{\sum_{h=2}^{NH}\frac{I_h^2}{2}}$ and $I_{rms} = \sqrt{I_{1rms}^2 + I_H^2}$. This last formulation is useful to describe the total harmonic distortion (THD) [20], which is a frequently used measure of harmonic levels. Mathematically, it is expressed as [21]:

$$THD_I = \frac{I_H}{I_{1rms}} \tag{6}$$

Another simple way to describe the harmonic influence over the fundamental frequency sinusoid is the distortion factor γ. Its formulation is as follows [21]

$$\gamma = \frac{I_{1rms}}{I_{rms}} = \frac{I_{1rms}}{\sqrt{I_{1rms}^2 + I_H^2}} \tag{7}$$

Solving for I_H in (6) and replacing it in (7):

$$\gamma = \frac{I_{1rms}}{\sqrt{I_{1rms}^2 + THD_I^2 I_{1rms}^2}} = \frac{1}{\sqrt{1 + THD_I^2}} \tag{8}$$

Later, by multiplying (6) and (7), the expression is:

$$THD_I\gamma = \frac{I_H}{I_{1rms}}\frac{I_{1rms}}{I_{rms}} = \frac{I_H}{I_{rms}} \tag{9}$$

By replacing (8) in (9) and solving for I_H, the relationship between the magnitude of the harmonic part and the total current magnitude is obtained:

$$I_H = \frac{THD_I}{\sqrt{1 + THD_I^2}} I = HI \tag{10}$$

where H is denoted the harmonic factor.

With respect to the vector representation of the phasor given in (2), the domains $\hat{i}$ and $\hat{j}$ correspond to the active (I_A) and reactive (I_X) current components, respectively. Nevertheless, in the presence of nonlinear loads, the harmonic component(I_H) appears in a third domain $\hat{k}$. This fact is attributed to the power vector configuration approach (applied for nonlinear loads) as presented in [22]. Then, the phasor in (5) can be written in component form as follows [17]:

$$\widetilde{I} = I_A\hat{i} + I_X\hat{j} + I_H\hat{k} \tag{11}$$

Then the apparent current magnitude is

$$I = \sqrt{I_A^2 + I_X^2 + I_H^2} \tag{12}$$

By replacing (10) into (12) and solving for I,

$$I = \sqrt{\frac{I_A^2 + I_X^2}{1 - H^2}} \tag{13}$$

The last formulation describes in functional way the mathematical model for a nonlinear load.

2.2 Electrical waveform disturbances

An electrical disturbance is characterized by the deviations that it produces to the nominal voltage, current, or frequency conditions. These fluctuations can result in failure or abnormal operation on the system. These perturbations can be noticed as wave deformations affecting magnitude or frequency mainly. This effect is of uttermost importance in electrical utilities since they face the task to provide high-quality energy by regulation, in addition to balance generation and demand with adequate levels of electromagnetic compatibility that allows proper operation of electrical equipment.

Some equipment with nonlinear components, such as power electronic converters, electric arc devices, and others, cause problems usually related to electromagnetic interference (EMI). These disturbances cause a loss of performance in most conventional loads and unnecessarily overload in transmission or distribution lines. However, one of the most significant problems in addition to the performance degradation is the deterioration on the quality of the voltage sine wave, superimposing periodic or transient disturbances. This phenomenon jeopardizes the appropriate operation of electronic, computer, and communication systems.

Given the aforementioned problems, there is a need to formulate a model that could handle analysis and simulation. **Table 2** shows the mathematical model and representation of the electrical disturbances analyzed in this chapter.

Perturbance	Representation	Mathematical model Assume $\boldsymbol{i(t)} = \boldsymbol{I_1} \cos(\omega_0 \boldsymbol{t} + \theta_1)$	
Momentary interruption	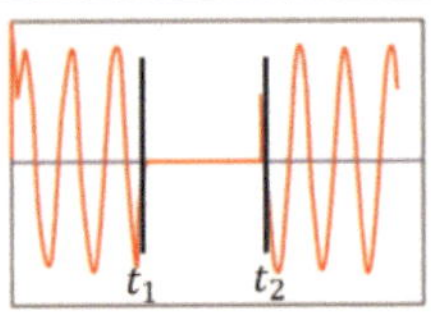	$i(t) = \begin{cases} I_1 \cos(\omega_0 t + \theta_1), t_1 \le t \\ 0 \quad , t_1 \le t \le t_2 \\ I_1 \cos(\omega_0 t + \theta_1), t \ge t_2 \end{cases}$ where $0.01 \le t \le 0.6$ s	(14)
Swell	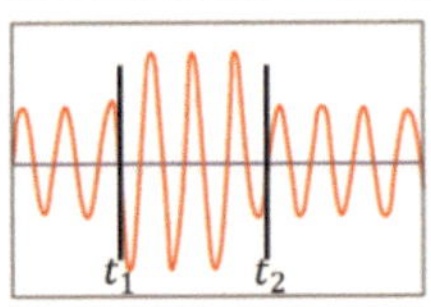	$i(t) = \begin{cases} I_1 \cos(\omega_0 t + \theta_1), t_1 \le t \\ \alpha I_1 \cos(\omega_0 t + \theta_1), t_1 \le t \le t_2 \\ I_1 \cos(\omega_0 t + \theta_1), t \ge t_2 \end{cases}$ where $\alpha>1$ and $0.01 \le t \le 0.6$ s	(15)
Sag	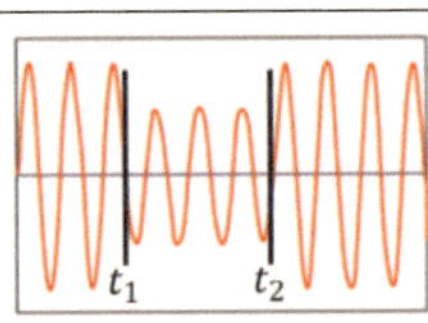	$i(t) = \begin{cases} I_1 \cos(\omega_0 t + \theta_1), t_1 \le t \\ \alpha I_1 \cos(\omega_0 t + \theta_1), t_1 \le t \le t_2 \\ I_1 \cos(\omega_0 t + \theta_1), t \ge t_2 \end{cases}$ where $0 < \alpha < 1$ and $0.01 \le t \le 0.6$ s	(16)
Harmonics	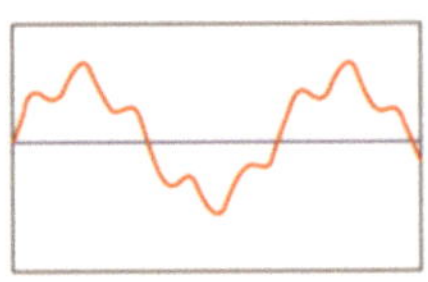	$i(t) = \sum_{h=1}^{NH} I_h \cos(h\omega_0 t + \theta_h)$	(17)

Table 2.
Electrical waveform disturbance mathematical model.

2.3 Solid-state transformer

The SST allows isolation between medium- and low-AC voltage sides as any conventional transformer. Additionally, it allows the isolation and clearance of faulty conditions from both sides, as well as anomalies encountered in the AC or DC sides. Its DC link is highly attractive for the integration of photovoltaic energy, storage systems with uninterrupted power supply devices, or even future local DC grids. In order to accomplish all these features, its topology has several stages of power electronic blocks depending on the functionalities required. Thus, the SST can be designed depending on the type of application [23]. As a key technology in the implementation of the smart grid, its topology will heavily depend on the end user consumption and the integration and coordination features required. Some of these requirements are shown in **Table 3**.

As the modular arrangement of the SST depends on the grid requirements, several topologies have been proposed in the literature. Generally, the energy can be processed in three main stages: rectification, the same level AC-AC or DC-DC conversion, and inversion. Some of the available solutions to these stages are shown in **Table 4**. To provide a wider classification system for the SST, the level of modularity can be determined with respect to power flow direction, connection to three-phase systems, and connection to the medium-voltage level [24].

Requirements		Description
Integration	Renewable energies	Integration of distributed generation on LVDC (e.g., photovoltaic panels) or LVAC (e.g., wind micro-turbines)
	Storage systems	Integration of energy storage system (e.g., battery systems) or devices with UPS functionality
Coordination	Power quality	Voltage magnitude (e.g., power factor correction) Reactive compensation (e.g., fast response to voltage disturbances due to reactive energy unbalances) Voltage unbalance (e.g., rapid response to sags, swells, and all the harmonics originated at the load or perceived at the device's input) Other quality events (e.g., electromagnetic transients, frequency variations)
	Remote operation	Communication functionalities to be integrated to higher management systems (e.g., SCADA systems, energy management systems (EMS), outage management systems (OTS), wide area management systems (WAMS), and other early awareness systems with synchro-phasor capabilities)
Consumption	Power supply	Several voltage level requirements: HVAC to LVAC, HVAC to LVAC + LVDC, etc. End user consumption such as LVAC loads (linear and nonlinear loads) and LVDC loads

Table 3.
Various functional requirements for the SST.

Rectification	Same level DC-DC conversion	Inversion
Full-bridge rectifiers	Buck/boost/buck-boost converter	Full-bridge inverters
Multilevel cascade rectifier	Cuk converter	
Active front-end rectifiers	Bi-directional DC-DC dual-active-bridge converter	

Table 4.
Typical power electronics topology for the SST stages.

A typical configuration of a SST consists of [25]:

1. Input filter, responsible for limiting the ripple of the input current, is composed of an inductor for each phase.

2. Full-bridge multilevel rectifier converts AC to DC voltage.

3. High-voltage DC (HVDC) link capacitors are energy storage for control purposes.

4. Dual active bridge (DAB) reduces the voltage level of each of the high links.

5. Low-voltage DC (LVDC) link capacitor is the link between the DAB and the three-phase inverter, which allows the integration of DC sources and DC loads.

6. Three-phase inverter converts the low link voltage into an alternating three-phase voltage.

7. LC filter is responsible for delivering an alternating sinusoidal voltage without distortion to the output of the SST.

To understand in a better way the SST topology, **Figure 3** is presented.

Focusing on the control system, it is divided into three stages: (1) multilevel cascade H bridge converter (MCHBC), which facilitates the conversion from AC to DC; (2) dual-active-bridge (DABC) DC-DC converter that allows the regulation of the energy in the low-voltage link capacitor and indirectly the DC voltage; (3) a three-phase inverter in series with a low-pass filter (inverter low-pass filter, ILP), which takes DC input wave and transforms it into AC without any distortion in the waveform. **Table 5** shows the description of the electrical circuit and mathematical model of each stage of the SST.

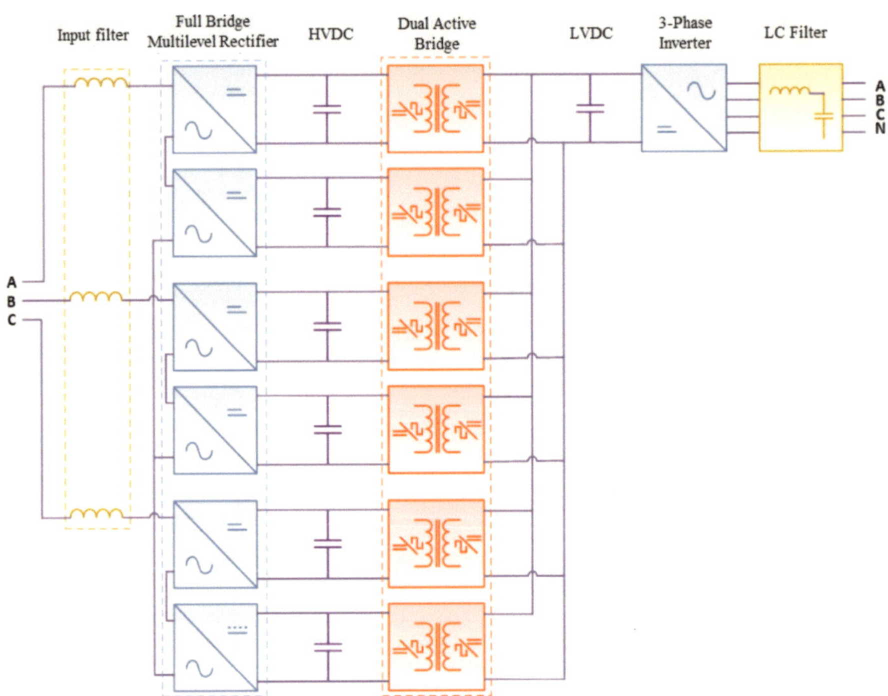

Figure 3.
Topology of the SST.

<table>
<tr><th>Stage</th><th>Electrical circuit</th><th>Mathematical model in frequency domain s</th></tr>
<tr><td>MCHBC</td><td>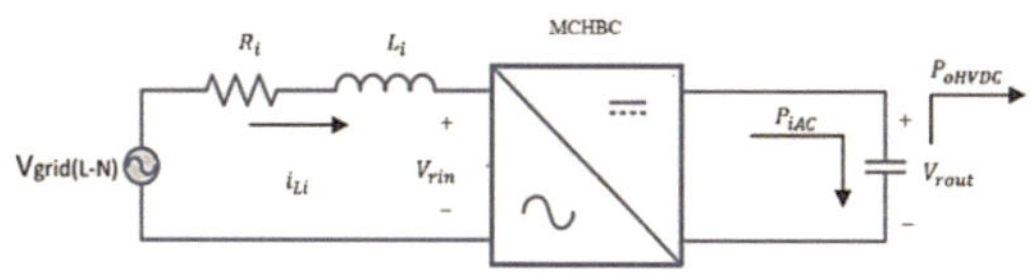
</td><td>$i_{Li}(s) = \frac{V_{grid}(s) - mV_{rout}(s)}{R_i + sL_i}$ (18)
$\frac{E_{CHVDC}(s)}{i_{Li}(s)} = \frac{V_{grid}(s)}{2s}$ (19)</td></tr>
<tr><td>DABC</td><td>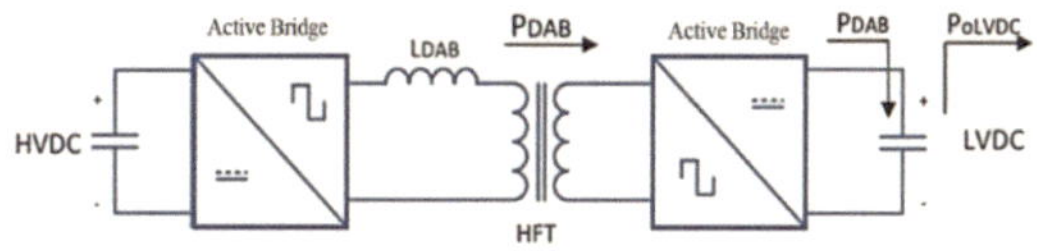
</td><td>$\frac{\Delta P_{DAB}}{\Delta \phi} = \frac{V_{HVDC} V_{LVDC}}{2\pi f_s L_{DAB}}$ (20)
$E_{CLVDC} = \frac{1}{s} P_{DAB}$ (21)</td></tr>
<tr><td>ILP</td><td>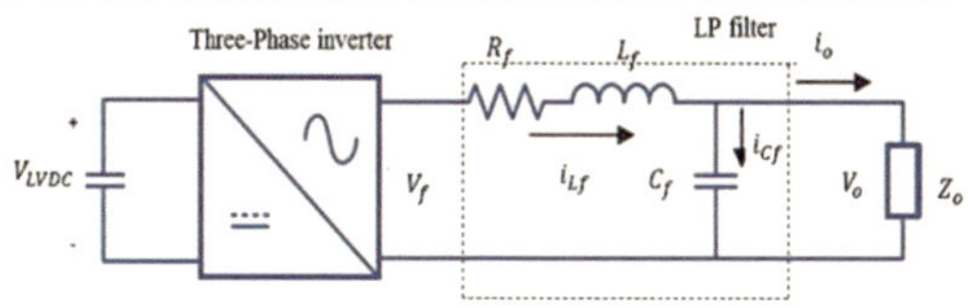
</td><td>$\frac{i_{Lf\ d}}{V_{f\ d} - V_{o\ d}} = \frac{1}{R_f + L_f s}$ (22)
$\frac{i_{Lf\ d}(s)}{V_{o\ d}(s)} = \frac{Z_o}{1 + sZ_o C_f}$ (23)
$\frac{i_{Lf\ q}}{V_{f\ q} - V_{o\ q}} = \frac{1}{R_f + L_f s}$ (24)
$\frac{i_{Lf\ q}(s)}{V_{o\ q}(s)} = \frac{Z_o}{1 + sZ_o C_f}$ (25)</td></tr>
</table>

Table 5.
Description of each mathematical model SST stage.

<table>
<tr><th>Stage</th><th>Driver</th></tr>
<tr><td>MCHBC</td><td>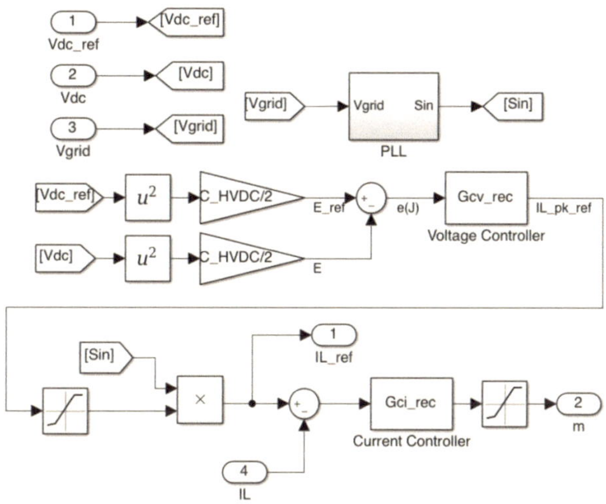
</td></tr>
<tr><td>DABC</td><td>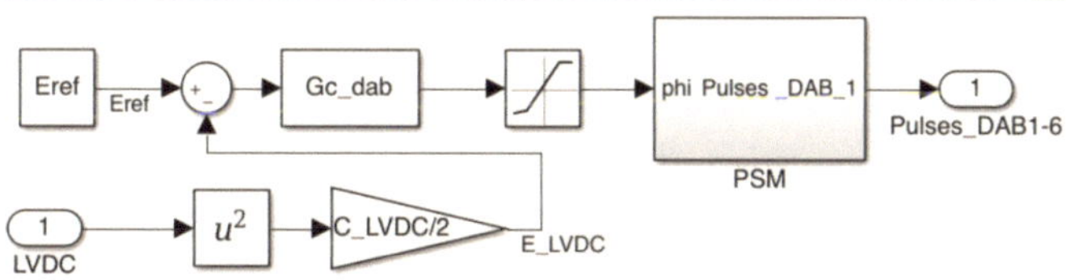
</td></tr>
<tr><td>ILP</td><td>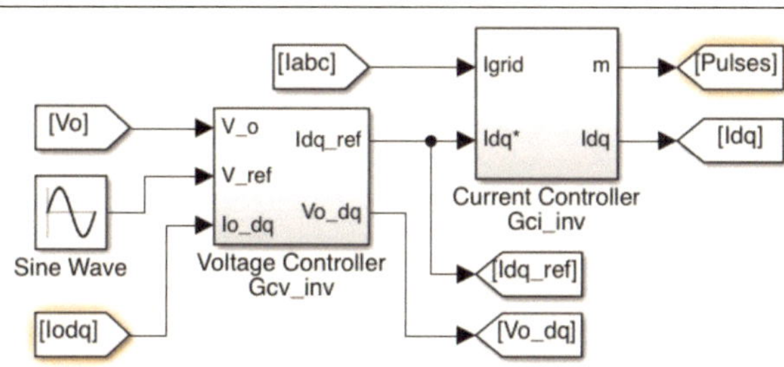
</td></tr>
</table>

Table 6.
Description of each controller SST stage.

The deduction of each formulation is presented in [26], where i_{Li} is the current flowing into the converter, V_{grid} is the voltage of the grid, m represents the modulation index, V_{rout} is the MCHBC voltage output, R_i is the resistance of the input filter, L_i is the inductance of the input filter, E_{CHVDC} is the energy storage in the high-voltage DC link, V_{LVDC} is the voltage in the low side of the transformer, V_{HVDC} is the voltage in the high side of the transformer, f_S is the switching frequency of the IGBTs, L_{DAB} is leakage inductance of the transformer, P_{DAB} is the power required by the system, ϕ is the phase angle between the high- and low-voltage side of the transformer, E_{CLVDC} is the stored energy in the capacitor, V_f is the voltage inverter, R_f is the filter resistance, L_f is the filter inductance, C_f is the filter capacitance, i_{Lf} is the current flowing out the inverter, V_o is the load voltage, and the terms $dq0$ represent the frames after applying Park transformation.

Once the SST mathematical formulation is defined, the drivers for each stage are designed. These are shown in **Table 6** [26].

3. Applications: energy enhancement

In this section, the SST is tested under different conditions. The analysis starts by describing the features of the SST, which are given in **Tables 7–9**.

3.1 Voltage sag

The grid is disturbed with a sag in the SST input. The sag appears with a voltage reduction of $30\% V_{grid}$ during a time of 0.05 s (between $0.15 \le t \le 0.20$). It is considered that there is a load of 400 kVA with a power factor of 0.85.

Input voltage	Output voltage	Power output	Voltage modulation index for each converter
13. 8 kV	440 V	800 kVA	0.85

Table 7.
SST nominal values.

MCHBC	DABC	ILP
$L_{A=B=C} = 41.9$ [mH]	$L_{DAB} = 1.29$ [mH]	$L_f = 286$ [mH]
$C_{HVDC} = 311$ [μF]	$C_{LVDC} = 585$ [μF]	$C_f = 22$ [μH]

Table 8.
Inductance and capacitance of each stage of the SST.

Stage	Control transfer functions
MCHBC	$G_{CI_{rec}} = \frac{-000{,}005{,}933\ s-1}{1276x10^{-9}s^2+00{,}002{,}988\ s}$
	$G_{CV_{rec}} = \frac{0,\ 0495\ s+1}{001{,}314\ s^2+3699\ s}$
DABC	$G_{C_{DAB}} = \frac{0,\ 001188\ s+1}{8739x10^{-6}s^2+01{,}025\ s}$
ILP	$G_{CI_{INV}} = \frac{0,\ 002376\ s+1}{9021x10^{-7}s^2+0{,}005{,}288\ s}$
	$G_{CV_{INV}} = \frac{1}{0{,}002{,}464\ s}$

Table 9.
Drivers of each stage of the SST.

Figure 4 reveals that even though the network voltage decreases (consequently the current injected also decreases), both the current and voltage on the load side are not affected. It is also observed that during the time the sag lasts, the inrush current increases. This increment is due to the SST control that keeps constant the output power, as shown in **Figure 5**.

The sag produces a decrement in the high DC voltage. To regulate it, the voltage modulation index (control) decreases. Its behavior is shown in **Figure 6**.

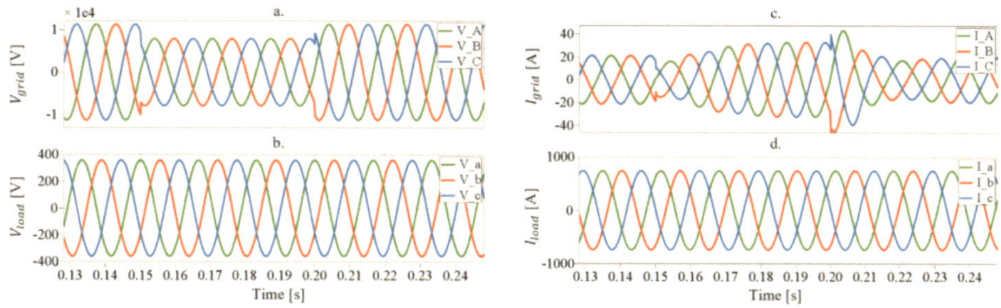

Figure 4.
Sag distortion waveform behavior: (a) grid voltage, (b) load voltage, (c) grid current, and (d) load current.

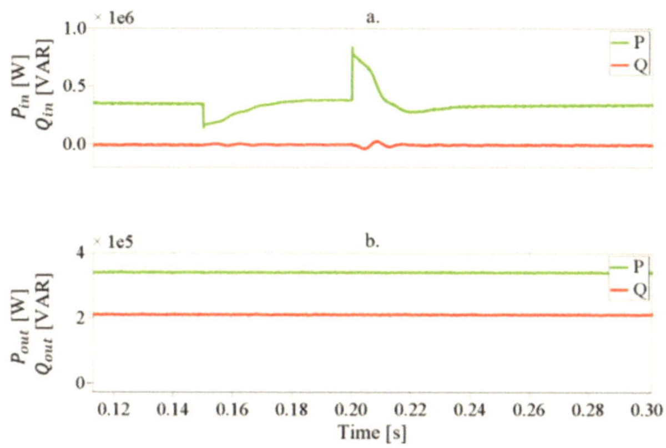

Figure 5.
Sag distortion power behavior at the SST (a) input and (b) output.

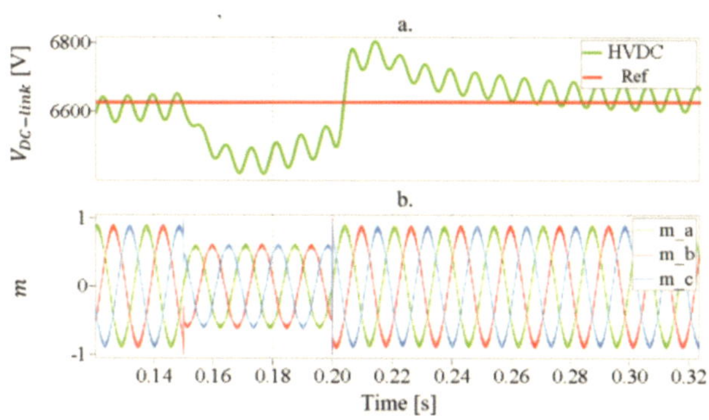

Figure 6.
Control response to sag disturbance: (a) HVDC and its reference and (b) modulation index.

3.2 Voltage swell

In this scenario the grid is disturbed with a swell in the SST input. The swell appears with a voltage increment of $15\% V_{grid}$ during a time of 0.05 s (between $0.15 \leq t \leq 0.20$). It is considered that there is a load of 800 kVA with a power factor of 0.90.

Figure 7 shows that although the swell disturbance at the SST input, the voltages and currents in the load side are not affected. It is also observed that during the time the swell lasts, the input current decreases. This is attributed to the control of the SST, which keeps constant the output power; this fact can be appreciated in **Figure 8**.

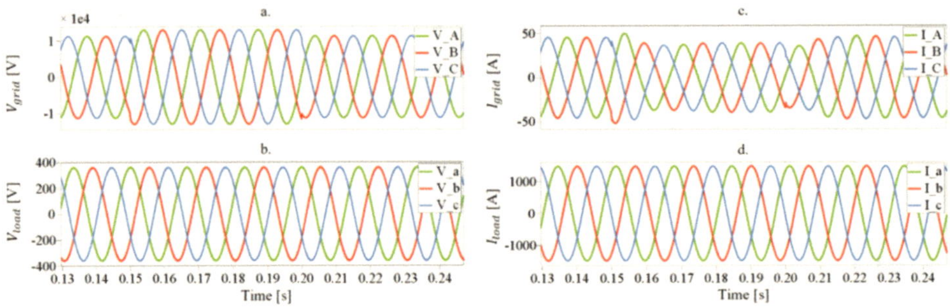

Figure 7.
Swell distortion waveform behavior: (a) grid voltage, (b) load voltage, (c) grid current, and (d) load current.

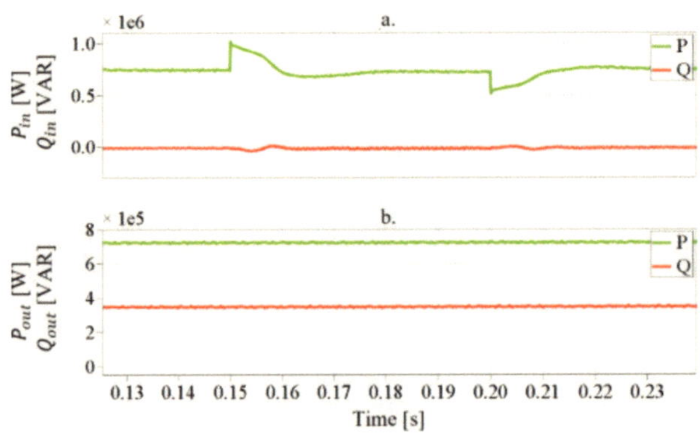

Figure 8.
Swell distortion power behavior at the SST (a) input and (b) output.

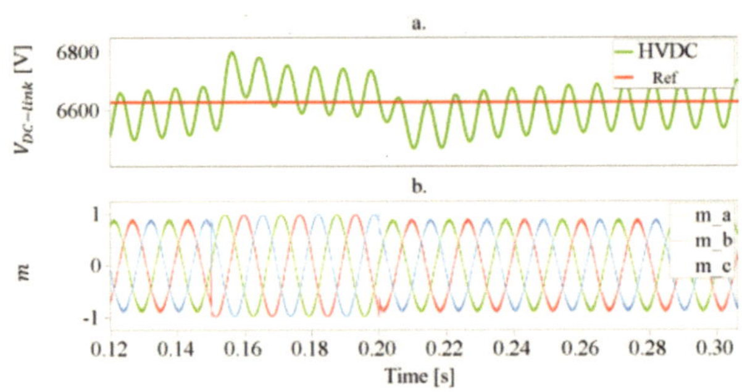

Figure 9.
Control response to swell disturbance: (a) HVDC and its reference and (b) modulation index.

The sag produces an increment in the high DC voltage. To regulate it, the voltage modulation index (control) increases. Its behavior is shown in **Figure 9**.

3.3 Harmonics by nonlinear loads

For this scenario, a nonlinear load of $z = 0.2711\hat{i} + 0.2763\hat{j}$ and $H = 0.233$ is connected to the SST. The harmonics generated are the third, fifth, and seventh. These harmonics have impact in voltage waveform of all the components connected into the grid [15] that means the generator will be also affected. Nevertheless, since the SST is connected, the harmonics are mitigated. In **Figure 10(a)**, the voltage waveform distortion produced by the nonlinear load is presented, and **Figure 10(b)** presents the voltage waveform of the grid, which shows no harmonic disturbances. In addition, it is observed in **Figure 11** that the harmonics in the harmonic currents have a negligible impact over the power flow in the grid side.

3.4 Overload and power factor

In this simulation, an R-L load of 0.7 power factor is connected. Initially, the load operates with a value of 500 kVA; then at t = 0.10 s, the load increases to 1000 kVA.

Under these conditions, it must be verified that the power factor at the input is approximately 1 and that the output voltage maintains its nominal value. **Figure 12**

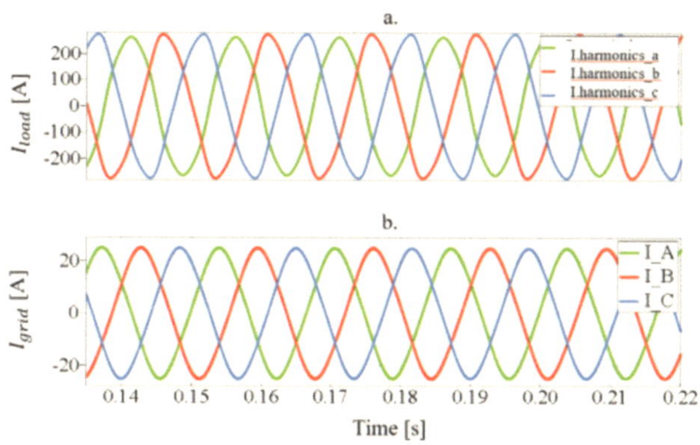

Figure 10.
Harmonic waveform behavior: (a) load current and (b) grid current.

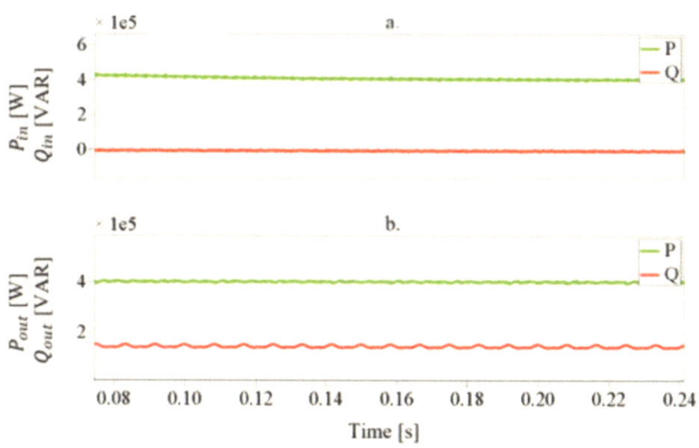

Figure 11.
Harmonic power behavior at the SST (a) input and (b) output.

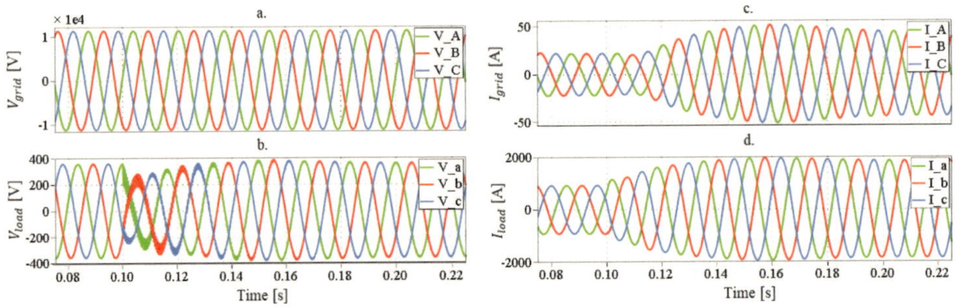

Figure 12.
Overload waveform behavior: (a) grid voltage, (b) load voltage, (c) grid current, and (d) load current.

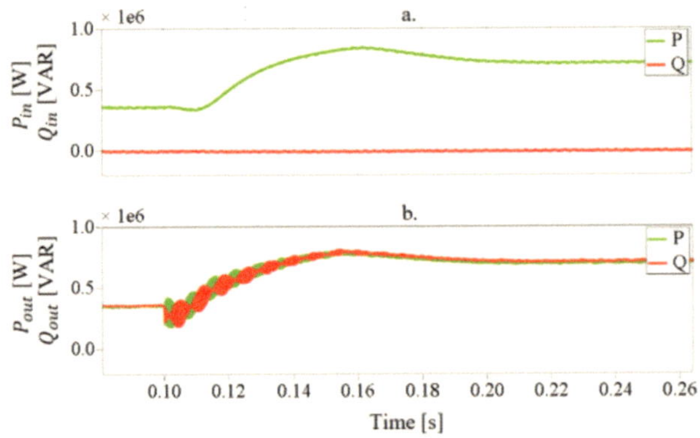

Figure 13.
Overload power behavior at the (a) grid side and (b) load side.

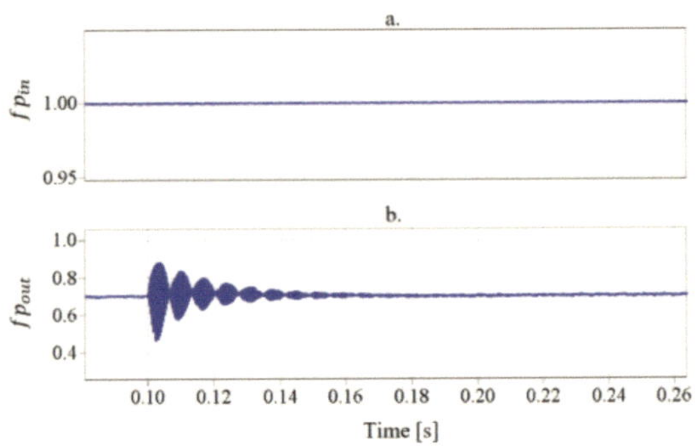

Figure 14.
Power factor behavior at the (a) grid side and (b) load side.

presents an increment in the magnitude voltage and a decrement in the current load; consequently the active and reactive power behavior is as given in **Figure 13**. The power factor in the load side does not affect the power factor at the input side (grid), as shown in **Figure 14**. It is verified that the SST can operate normally with an overload of 125%, and the power factor improves.

3.5 Bi-directional power flow

In this scenario, a distributed generation and an energy storage are connected to the DC link of the SST, with a voltage operation of 1144 V as presented in **Figure 15**.

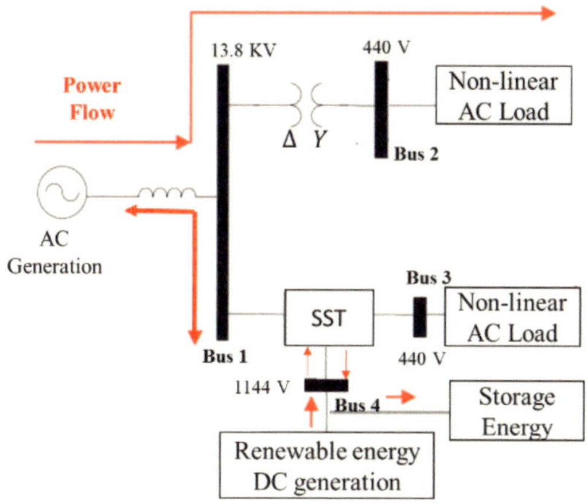

Figure 15.
Electrical diagram for a bi-directional power flow.

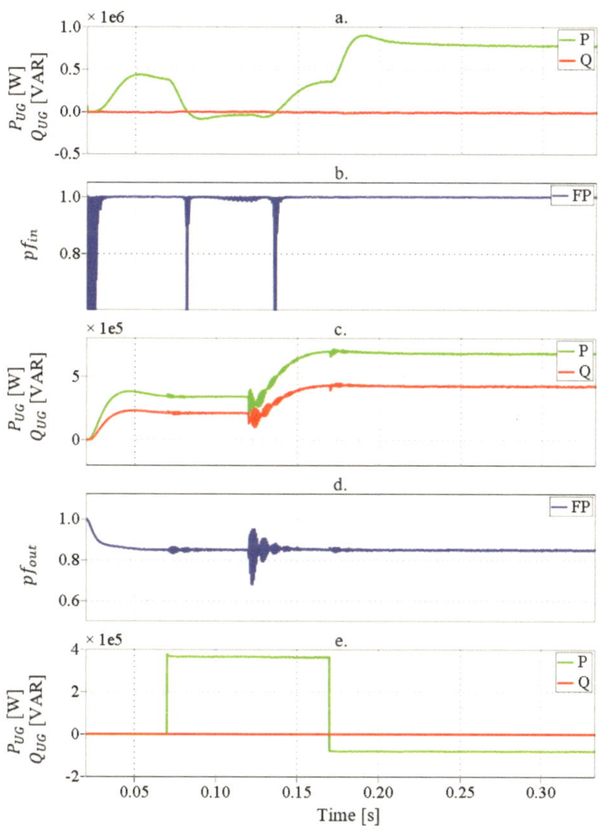

Figure 16.
(a) Generator active and reactive power, (b) generator power factor, (c) R-L load power factor, (d) power factor of the distributed energy source, and (e) power of the distributed energy source and the storage energy.

Initially, a load of 50% of their nominal demand is connected; later the load increases to 100% with a power factor of 0.85 lagging, as shown in **Figure 16(c)**.

It is observed that at $t = 0.07$ s, the distributed generation starts to deliver active power, as shown in **Figure 16(d)**. Then the input power coming from the grid decreases until reaching a negative value as presented in **Figure 16(a)**; this means that the distributed generation is delivering power to the load and the grid. Then at

$t = 0.17$ s, the load increases its demand; that means, there is more power consumption by the load, and this causes the grid to start delivering power to the load. The simulation continues in such a way that the distributed generation is switched off and instead the storage energy starts operating.

4. Communication requirements

It is possible to deploy a communication system that could satisfy the communication requirements and provide an enhanced operational capability in an SST

Technology		**Data range**	**Range**	**Use in smart grid**
Power line communication (PLC)	Narrowband PLC (NB-PLC)	NB-PLC: 1–10 Kbps (low data rate PHYs) 10–500 Kbps (high data rate PHYs)	NB-PLC: ~150 km or more	NB-PLC: Large-scale automatic metering infrastructure (AMI) NAN/FAN WAN
	Broadband PLC (BB-PLC)	BB-PLC: 1–10 Mbps (long range) ~200 Mbps (short range)	BB-PLC: ~1.5 km	BB-PLC: Small-scale AMI HAN
Optical communications	Active optical networks (AON)	AON: 100 Mbps (up/down)	AON: ~10 km	WAN NAN/FAN AMI (with FTTH systems)
	Passive optical networks (PON): BPON, -EPON GPON	PON BPON 155–622 Mbps (up/down) GPON: 155–2448 Mbps (up) 1.244–2.448 Gbps (down). EPON: 1 Gbps up/(down)	BPON, GPON: ~20–60 km EPON: ~10–20 km	
Digital subscriber line (DSL)	ADSL	ADSL: 8 Mbps (down) and 1.3 Mbps (up) ADSL2: 12 Mbps (down) and up to 3.5 Mbps (up). ADSL2+: 24 Mbps (down) and up to 3.3 Mbps (up)	ADSL: ~4 km ADSL2: ~7 km ADSL2+: ~7 km	AMI NAN/FAN
	VDSL	VDSL: 52–85 Mbps (down and 16–85 Mbps (up) VDSL2: up to 200 Mbps (down/up)	VDSL: ~1.2 km VDSL2: ~300 m (maximum rate)–1 km (50 Mbps)	

Table 10.
Wired technologies for SG.

network. Several types of topologies can be considered depending on the application. For instance, in a star-type topology, the communication linkage is established between each SSTs and the control center directly. Other topologies allow improved connectivity with alternate connections and meshed links. However, in all cases, a certain level of security, scalability, and minor delay in the information and bi-directional data transfer capabilities is required. While information capability performs digital monitoring of SST variables (as in SCADA systems), the bi-directional data transfer capability allows fast responses to disturbances such that system's performance can be improved accordingly [27]. In fact, the smart grid (SG) concept is based on reliable real-time data availability and utilization for more intelligent decision-making.

There could be two forms of communication in SST networks: wired or wireless. Their selection depends on the bandwidth and the cost of the telecommunications infrastructure [28]. In the wired case, there are technologies based on power line communications (PLC) and optical communications and digital subscriber line (DSL) [29]. **Table 10** shows the comparison of wired communication technologies for smart grids according to coverage range and maximum theoretical data transmission. It is observed that optical fiber main application is the connectivity between transmission/distribution substations, thus, forming large coverage areas satisfying very high volumes of data and low latency. However, the main disadvantage is its high installation and equipment costs. On the other hand, PLC and DSL are technologies that can be merged on existing copper-wired networks, but their bigger limitations are scalability and network flexibility [30].

In the case where the installation is above ground level, SSTs could have a wireless communication system. In fact, whenever possible, wireless technologies are preferred due to its flexibility and low cost; they can cover difficult access areas (distant or inaccessible) in power system monitoring applications [31]. As an example, a multipoint to point (MP2P) communication system for SST-based power system is shown in **Figure 17**. There are several wireless technologies that depend on the coverage and data rate, and these technologies allow the adoption of the multilayer architecture for smart grid as shown in **Table 10**. In the case of an SST-control center communication network, it is also possible to incorporate different intelligent electronic devices (IEDs), remote terminal units (RTU), substation automation solutions (SAS), universal gateways, smart meters, etc. There will be an increased complexity in the network operation due to the large amounts of data. Hence, these types of applications will require higher reliabilities and lower latencies.

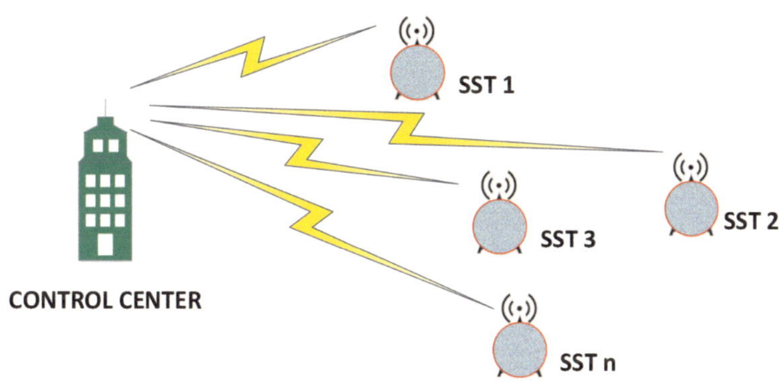

Figure 17.
Communication network for SST.

There are several wireless technologies that depend on the coverage and data rate, and these technologies allow the adoption of the multilayer architecture for smart grid as shown in **Table 10**. In the case of an SST-control center communication network, it is also possible to incorporate different intelligent electronic devices (IEDs), remote terminal units (RTU), substation automation solutions (SAS), universal gateways, smart meters, etc. There will be an increased complexity in the network operation due to the large amounts of data. Hence, these types of applications will require higher reliabilities and lower latencies. For such complex networks, a geographical-dependent structure is required.

According to the geographical service, networks are classified in home area network (HAN), neighborhood area networks (NANs), and wide area network (WAN). These networks have different coverage areas as detailed in **Figure 18**. HAN refers to networks within a single point facility (e.g., substation); it can range from a single home to a business area network (BAN) or industrial area network (IAN). Outside the single point facility, there are NANs and WANs. NAN, also known as field area network (FAN), connects several HANS and covers the transmission or distribution areas within several square kilometers. On the other hand, WAN connects several NANs, and it is considered the backbone of the communication system. It can cover thousands of square kilometers including the main control center. WAN can be a hybrid network with a mixture of wired and wireless sections [32].

For applications that could be deployed wirelessly, the reader can find an updated selection of available technologies including satellite and mobile communications in **Table 11**. The communication spectrum could present congestion in

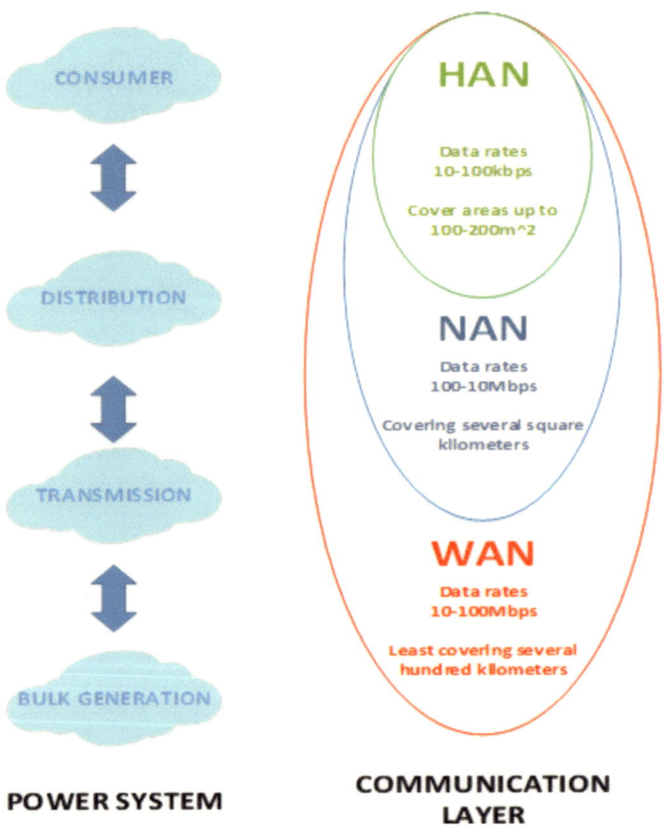

Figure 18.
Communication layer for SST.

Technology		Data rate	Range	Use in smart grid
WPAN IEEE 802.15		256 Kbps	Between 10 and 75 m	Vehicle-to-grid (V2G) HAN: AMI
Wi-Fi	IEEE802.11e (QoS-enhancements)	IEEE 802.11e/s: ~54 Mbps	IEEE 802.11e/s/n: ~300 m (outdoors)	V2G HAN AMI
	IEEE802.11n (ultrahigh network throughput)	IEEE 802.11 n: ~600 Mbps		
	IEEE802.11 s (mesh networking)	IEEE 802.11af: ~26.7 Mbps		
	IEEE802.11p wireless access in vehicular environments (WAVE)	IEEE 802.11ah: ~40 Mbps	IEEE 802.11p: ~1 km IEEE 802.11ah: ~1 km IEEE 802.11af: >1 km	
WiMAX	IEEE 802.16 (fixed and mobile broadband wireless access)	IEEE802.16: 128 Mbps down and 28 Mbps up	IEEE 802.16: 0–10 km	AMI NAN/FAN WAN
	IEEE 802.16j (multi-hop relay)	IEEE802.16 m: 100 Mbps for mobile users, 1 Gbps for fixed users	IEEE 802.16 m: 0–5 km (optimum) 5–30 km (acceptable) 30–100 (reduced performance) km	AMI NAN/FAN WAN
	IEEE802.16 m (advanced air interface)			
Cellular communications 3G	HSPA	14.4 Mbps down and 5.75 Mbps up	HSPA+: 0–5 km	V2G HAN: AMI NAN WAN
	HSPA+	84 Mbps down and 22 Mbps up		
Cellular communications 4G	LTE	326 Mbps down and 86 Mbps up	LTE-Advanced: 0–5 km (optimum) 5–30 km (acceptable) 30–100 km (reduced performance)	
	LTE-advanced	1 Gbps down and 500 Mbps up		
Satellite	LEO	Iridium: 2.4–28 Kbps Inmarsat-B: 9.6 up to 128 Kbps BGAN: 384 up to 450 Kbps	Depend on number of satellites and their beams	WAN AMI

Table 11.
Wireless technologies for SG.

licensed and unlicensed bandwidths due to the increasing number of technologies sharing the same resource. Therefore, the network designer must consider more stringent security mechanisms. A more efficient spectrum can deliver increased

Smart metering and grid applications		Customer applications	Application layer
Authentication, access control, integrity protection, encryption, privacy			Security layer
Cellular, WiMAX, optical fiber	PLC, DSL, IEEE 802.22	Wi-fi, ZigBee, Bluetooth	Communication layer
WAN	NAN/FAN	HAN/BAN/IAN	
PMUs Cap bank Reclosers	Switches Sensors	Transformers Meters Storage	Power control layer
Power transmission and generation	Power distribution	Customer premises	Power system layer

Table 12.
System multilayer architecture of SG [30].

Quantitative requirements	Qualitative requirements
Latency Reliability Data rate	Scalability Interoperability Flexibility Security Regulatory issues

Table 13.
Network requirements for SST over SG.

data rates and provide enhanced interoperability between devices and systems, as shown in the system architectures of **Table 12**. The main features for an efficient communication can be established through several qualitative and quantitative requirements for the SST-based power system telecommunications infrastructure, as shown in **Table 13**. It is important to highlight that many of the technologies of **Tables 10** and **11** are integral in today's power system operation, such as the advanced metering infrastructure (AMI), energy management system (EMS), wide area management systems (WAMS), etc. For the case of the SST-based power system, the wired and wireless technologies could provide a systemic integration and seamless communication (**Table 12**).

5. Conclusion

The voltage supply should ideally have a waveform without deformations. However, nonlinear loads produce voltage waveform distortion that affects the quality of the grid, leading to a low energy efficiency. It is not possible to mitigate their presence since they have become part of daily life. Nonetheless, the implementation of smart devices (such as the SST) can hold on its effects, becoming in a potential solution to this problem.

The presence of SST in a power system can improve the power quality of the grid. The SST allows to uncouple the side of the network from the side of the load; then if a disturbance occurs from one side, it does not affect the components connected in the other side of the SST. In addition, the SST allows to enhance the power factor, support overloads, and keep nominal voltage on the load side, even though the input voltage is affected by either a sag or a swell. Another advantage of the SST is their DC link, which allows the integration of distributed generation and energy storage. The power coming from the DC link can deliver power to the network, if required.

Concerning the communication, the SST faces a great challenge. The requirements for SST's wireless communication network are complex because they seek lower latency, greater bandwidth, interoperability, and scalability. For this reason, it is relevant to focus on involving other types of wireless networks as an alternative solution.

Acknowledgements

This study was supported by Escuela Superior Politécnica del Litoral (ESPOL), the Electrical System Research Group GISE of the Faculty of Electrical Engineering and Computer Science FIEC (ESPOL), the scholarship program Walter Valdano Raffo II (ESPOL), and the Secretariat of Higher Education, Science, Technology and Innovation of the Republic of Ecuador (SENESCYT).

Conflict of interest

No potential conflict of interest is reported by the authors.

Author details

Fernando Vaca-Urbano*, Manuel S. Alvarez-Alvarado, Angel A. Recalde and Félix Moncayo-Rea
ESPOL Polytechnic University, Escuela Superior Politécnica del Litoral ESPOL, Facultad de Ingeniería en Electricidad y Computación, Guayaquil, Ecuador

*Address all correspondence to: fearvaca@espol.edu.ec

References

[1] Chakraborty A. Advancements in power electronics and drives in interface with growing renewable energy resources. Renewable and Sustainable Energy Reviews. 2011;**15**(4):1816-1827

[2] Colak I, Kabalci E, Fulli G, Lazarou S. A survey on the contributions of power electronics to smart grid systems. Renewable and Sustainable Energy Reviews. 2015;**47**:562-579

[3] Pepermans G, Driesen J, Haeseldonckx D, Belmans R, D'haeseleer W. Distributed generation: Definition, benefits and issues. Energy Policy. 2005;**33**(6):787-798

[4] Lopes JAP, Hatziargyriou N, Mutale J, Djapic P, Jenkins N. Integrating distributed generation into electric power systems: A review of drivers, challenges and opportunities. Electric Power Systems Research. 2007;**77**(9): 1189-1203

[5] Recalde AA, Saha TK, Mosadeghy M. Reliability evaluation with wind turbines and photovoltaic panels. In: Transmission & Distribution Conference and Exposition-Latin America (PES T&D-LA); IEEE PES; 2014. pp. 1-5

[6] Alvarez-Alvarado MS, Jayaweera D. Reliability model for a static var compensator. In: 2017 IEEE Second Ecuador Technical Chapters Meeting (ETCM); 2017. pp. 1-6

[7] Alvarez-Alvarado MS, Jayaweera D. A new approach for reliability assessment of a static var compensator integrated smart grid. In: 2018 IEEE International Conference on Probabilistic Methods Applied to Power Systems (PMAPS); 2018. pp. 1-7

[8] Alvarez-Alvarado MS, Jayaweera D. Aging reliability model for generation adequacy. In: 2018 IEEE International Conference on Probabilistic Methods Applied to Power Systems (PMAPS); 2018. pp. 1-6

[9] Rodríguez-Gallegos CD, Alvarez-Alvarado MS. Understanding the stationary and transient state of a solar array: Model and simulation. Latin-American Journal of Physics Education Volume. 2015;**9**(1):1501-1508

[10] Gungor VC et al. Smart grid technologies: Communication technologies and standards. IEEE Transactions on Industrial Informatics. 2011;**7**(4):529-539

[11] Fang X, Misra S, Xue G, Yang D. Smart grid—The new and improved power grid: A survey. IEEE Communication Surveys and Tutorials. 2012;**14**(4):944-980

[12] Gungor VC et al. A survey on smart grid potential applications and communication requirements. IEEE Transactions on Industrial Informatics. 2013;**9**(1):28-42

[13] Ipakchi A, Albuyeh F. Grid of the future. IEEE Power and Energy Magazine. 2009;**7**(2):52-62

[14] Popović-Gerber J et al. Power electronics enabling efficient energy usage: Energy savings potential and technological challenges. IEEE Transactions on Power Electronics. 2012;**27**(5):2338-2353

[15] Shah DG, Crow ML. Stability design criteria for distribution systems with solid-state transformers. IEEE Transactions on Power Delivery. 2014; **29**(6):2588-2595

[16] Burch R et al. Impact of aggregate linear load modeling on harmonic analysis: A comparison of common practice and analytical models. IEEE

Transactions on Power Delivery. 2003; **18**(2):625-630

[17] Alvarez-Alvarado M, Rodríguez-Gallegos C, Jayaweera D. Optimal planning and operation of static VAR compensators in a distribution system with non-linear loads. IET Generation Transmission and Distribution. 2018;**12**: 3726-3735

[18] Glover JD, Sarma MS, Overbye T. Power System Analysis & Design, SI Version. USA: Cengage Learning; 2012

[19] Singh B, Chandra A, Al-Haddad K. Power Quality: Problems and Mitigation Techniques. UK: John Wiley & Sons; 2014

[20] Grady WM, Gilleskie RJ. Harmonics and how they relate to power factor. In: EPRI Proceeding of the Power Quality Issues & Opportunities Conference; 1993. pp. 1e-8e

[21] Akagi H, Watanabe EH, Aredes M. Instantaneous Power Theory and Applications to Power Conditioning. Vol. 62. UK: John Wiley & Sons; 2017

[22] Hoevenaars A. How Harmonics Have Contributed to Many Power Factor Misconceptions. Canada: Mirus Int. Inc; 2014

[23] She X, Huang AQ, Burgos R. Review of solid-state transformer technologies and their application in power distribution systems. IEEE Journal of Emerging and Selected Topics in Power Electronics. 2013;**1**(3):186-198

[24] Kolar JW, Ortiz G. Solid-state-transformers: Key components of future traction and smart grid systems. In: Proceedings of the International Power Electronics Conference (IPEC), Hiroshima, Japan; 2014

[25] Falcones S, Ayyanar R, Mao X. A DC–DC multiport-converter-based solid-state transformer integrating distributed generation and storage. IEEE Transactions on Power Electronics. 2013;**28**(5):2192-2203

[26] Vaca-Urbano F, Alvarez-Alvarado MS. Power quality with solid state transformer integrated smart-grids. In: Innovative Smart Grid Technologies Conference-Latin America (ISGT Latin America); 2017 IEEE PES; 2017. pp. 1-6

[27] IEEE Guide for the Interoperability of Energy Storage Systems Integrated with the Electric Power Infrastructure. IEEE P2030.2/D9.0; December 2014; 2015. pp. 1-136

[28] Parikh PP, Kanabar MG, Sidhu TS. Opportunities and challenges of wireless communication technologies for smart grid applications. In: Power and Energy Society General Meeting; 2010 IEEE; 2010. pp. 1-7

[29] Ancillotti E, Bruno R, Conti M. The role of communication systems in smart grids: Architectures, technical solutions and research challenges. Computer Communications. 2013;**36**(17–18): 1665-1697

[30] Kuzlu M, Pipattanasomporn M, Rahman S. Communication network requirements for major smart grid applications in HAN, NAN and WAN. Computer Networks. 2014;**67**:74-88

[31] Alam S, Sohail MF, Ghauri SA, Qureshi IM, Aqdas N. Cognitive radio based smart grid communication network. Renewable and Sustainable Energy Reviews. 2017;**72**:535-548

[32] Nafi NS, Ahmed K, Gregory MA, Datta M. A survey of smart grid architectures, applications, benefits and standardization. Journal of Network and Computer Applications. 2016;**76**:23-36

Chapter 8

Hybrid Modeling Procedure of Li-Ion Battery Modules for Reproducing Wide Frequency Applications in Electric Systems

Sandra Castano-Solis, Daniel Serrano-Jiménez, Jesús Fraile-Ardanuy, David Jiménez-Bermejo and Javier Sanz-Feito

Abstract

In this chapter, a hybrid modeling procedure of Li-ion battery modules is presented. From experimental results, the parameters of an electrical circuit have been determined by means of time- and frequency-domain tests. In this way, the dynamic behavior of the battery-pack is modeled. The tests have been performed at the whole battery-pack, instead of a single-cell approach, in order to consider the packaging effects of multicell devices. The real performance of the battery-pack under dynamic applications associated with distribution grids has been simulated using a hardware-in-the-loop (HIL) experimental setup. According to simulation results, the hybrid model follows the battery-pack response with high accuracy.

Keywords: battery-pack modeling, dynamic performance electric grids, time-domain test, frequency-domain tests, HIL simulations

1. Introduction

Introducing renewable energy sources such as photovoltaic generators and wind turbines into energy distribution grids presents some drawbacks because the energy generation is discontinuous and strongly depends on daily weather conditions. For these reasons, distribution grids with high penetration of renewable resources present problems of reliability, stability, and power quality [1]. To solve these issues, several researchers recommend energy storage systems as support systems [2, 3].

In recent years, the improvements in terms of materials, high specific energy and power, and long life cycle have made Li-based batteries as a viable option to reduce renewable generation interruptions [4]. Because the nominal voltage of Li-ion cells is less than 4 V, commercial devices are composed by several Li-ion cells in a combination of series and/or parallel connections, to provide the desired power and capacity of the grid-scale applications. Multicell Li-ion devices include a battery management system (BMS) in order to prevent the cells' voltage, temperature, and charging/discharging current from exceeding the safety limits [5]. Also, the BMS uses algorithms to equalize the cells' voltage to avoid the cells' voltage differences (produced by manufacturing processes and/or ageing processes) from limiting the whole pack performance [6]. Besides electrochemical behavior of Li-ion cells, the performance of the battery-pack is affected by the interactions between cells and BMS components. This configuration increases the nonlinear behavior of commercial battery devices compared with single cells. For all these aforementioned reasons, modeling of battery-packs is a difficult task.

Several battery models have been proposed in literature in order to facilitate their integration into different applications. The most detailed models include electrochemical- or physical-based models, which are able to accurately describe the chemical processes taking place inside battery cells [7, 8]. Despite their accuracy, these models are very complex to be implemented in a simulation tool, and the coupled nonlinear differential equations that compose the model require heavy computational work [9, 10]. In contrast with electrochemical models, electrical circuit models are not very complex, allow the simulation of the electrical response of the battery by using electrical elements (resistances, RC networks, ideal voltage sources, etc.), and can be easily incorporated in control strategies and simulation platforms. The simplest electrical circuit model of a battery is given by Thevenin's equivalent circuit, which is composed of an ideal voltage source in series with a constant internal resistance [11–13]. This model can be used in an initial stage of battery dimensioning, but in the case of dynamic applications, it does not offer information regarding the transient behavior of battery [10, 14].

The accuracy of a battery model depends on the procedure used to obtain its parameters. Modeling techniques can be classified from simple black-box approaches to time- or frequency-domain procedures. Black-box models are simple to obtain, but do not provide information on the battery's internal behavior. Time-domain models are obtained from the analysis of the battery voltage evolution during charge–discharge tests by means of the procedure called current interruption test. In order to improve the model accuracy, some authors [15, 16] use online parameter identification methods to predict battery dynamical behavior as a function of time. These models are relatively easy to obtain, but their validity is usually limited to specific load regimes [17]. Frequency-domain based models are performed by means of electrochemical impedance spectroscopy (EIS) tests [18, 19]. In this technique, a small AC excitation signal (either current or voltage) is applied as a variable frequency sweep to the battery. To obtain a linear model, the amplitude of the AC excitation signal applied to the cell is kept between 5 and 10% of the rated voltage/current. Although EIS models can be time-consuming to obtain, they can reproduce accurate battery behavior in a wide frequency range, typically from mHz to kHz [20, 21]. The equivalent complex impedance is calculated as the quotient between the instantaneous values of voltage and current for each test point.

Li-ion battery-packs are modeled in the majority of the cases as an aggregation of individual cell models, neglecting the packaging effects of multicell devices, although some recent works have shown that considering the interactions between cells and BMS elements can improve the accuracy of the Li-ion battery-packs models [22, 23]. Also, most battery models do not consider that battery modules can work at different dynamic regimes due to internal electrochemical processes that affect their transient behavior. In highly dynamic applications such as electrical grid support or frequency control in microgrids, there are three time constants of special interest [24]. The first one corresponds to the fast processes with dynamic performance from millisecond to seconds, and it is related to safety control of the battery-pack. The second constant refers to the load regime, which produces different charging/discharging cycles of the battery-pack. Finally, ageing processes that occur during long time (months or years) affect the state of charge (SOC) estimation. All these aspects should be taken into account in the modeling procedure of Li-ion battery-packs in order to reproduce their real behavior.

This work presents a hybrid modeling procedure of battery-packs based on time- and frequency-domain tests. From experimental results, the parameters of an electrical circuit are calculated. The elements of the electric circuit are a voltage source, which is determined by current interruption tests (in time domain), and an impedance measured by electrochemical impedance spectroscopy tests (in frequency domain). All tests have been carried out at the whole battery assembly, instead of single-cell measurements, in order to consider the packaging effects of multicell devices. The model has been experimentally validated using hardware-in-the-loop (HIL) simulations. In this way, the battery-pack performance under high dynamic load regimes in distribution grids has been reproduced.

The chapter's contents are organized as follows: Section 2 explains the proposed hybrid model procedure, Section 3 presents the validation tests using a hardware-in-the-loop experimental setup, and finally, in Section 4, the conclusions are presented.

2. Hybrid modeling procedure

To reproduce the behavior of Li-ion batteries using an electric circuit, the circuit topology includes a voltage source that represents the active behavior of the battery and a series impedance of the passive one. In the hybrid experimental procedure proposed in this work, the parameters of the electrical model have been calculated from experimental results of both time- and frequency-domain tests.

The modeling procedure is applied to a commercial battery-pack composed of four parallel-connected strings (seven cells connected in series in each string) and a battery management system. BMS functions include measurement of cell voltage, temperature, and current of each series connection, an algorithm for cell voltage balancing, and the disconnection during charging and discharge processes (to avoid cells over/under voltage). **Figure 1** shows a battery-pack connection scheme. The technical data of the battery-pack are presented in **Table 1**.

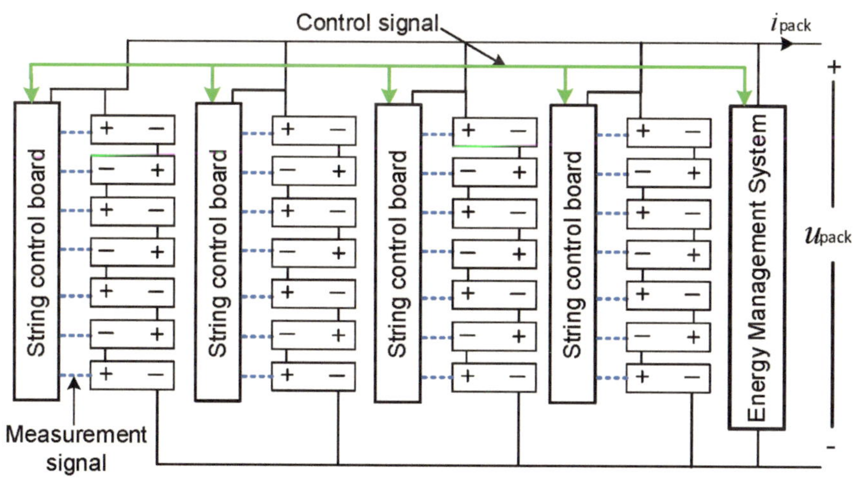

Figure 1.
Battery-pack connections.

Cells reference	MP 176065 Int (Saft batteries)
Pack rated voltage	25.9 V
Pack maximum voltage	29.4 V (4.2 V per cell)
Pack minimum cutoff voltage	20.3 V (2.9 V per cell)
Pack capacity, C_n	50 Ah
Pack maximum current	50 A
Range of temperature (charge)	−20 to 60°C
Range of temperature (discharge)	−30 to 55°C

Table 1.
Battery-pack technical data.

2.1 Modeling of the voltage source

To model the voltage source (E_0) of the battery-pack, the relationship between the open circuit voltage (OCV) and the state of charge is calculated by means of current interruption tests. First, the battery-pack is totally charged applying the constant current-constant voltage method (25 A until reaching the maximum voltage). After charging process, the device is discharged at current pulses of 10 A for 30 min (0.1 SOC variation) followed by 90 min of relaxation time. The OCV value for each test point is recorded when the relaxation time ends. Finally, the battery-pack is recharged at current pulses of 10 A for 30 min (0.1 SOC variation) followed by 90 min of relaxation time as in the case of discharge process. In the same manner, the OCV values are recorded at the end of the relaxation time. The results of these tests are shown in **Figure 2** (discharging test) and **Figure 3** (charging test). **Table 2** presents the OCV values associated with each SOC test point, and this relationship is also sketched in **Figure 4**. As it can be seen, at the end of charge test, the final value of 100% of SOC is not reached because the BMS limits the applied current during the two last pulses. It is important to highlight that this situation does not occur when a single cell is tested, because the cell is charged and discharged from 100% SOC to 0% SOC without protection of BMS. In addition, test results do not show a high deviation of the average values as is reported in literature [25]; therefore, these values are used to evaluate the OCV-SOC relationship, which is presented in Eq. (1).

$$E_o(SOC, t) = 26.05 - 0.15 \cdot SOC + 3.51 \cdot SOC^2 \quad (1)$$

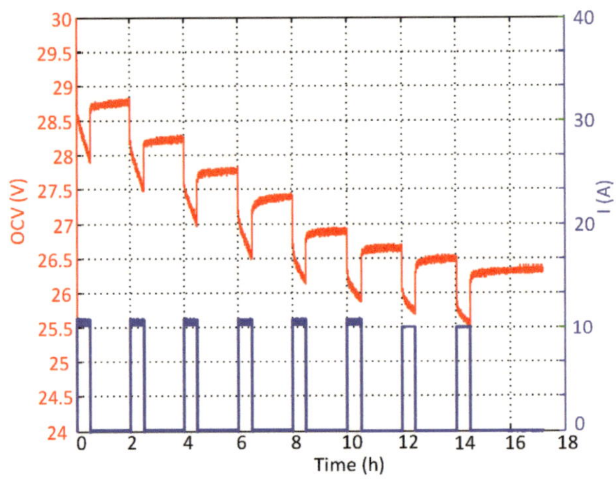

Figure 2.
Discharge test result.

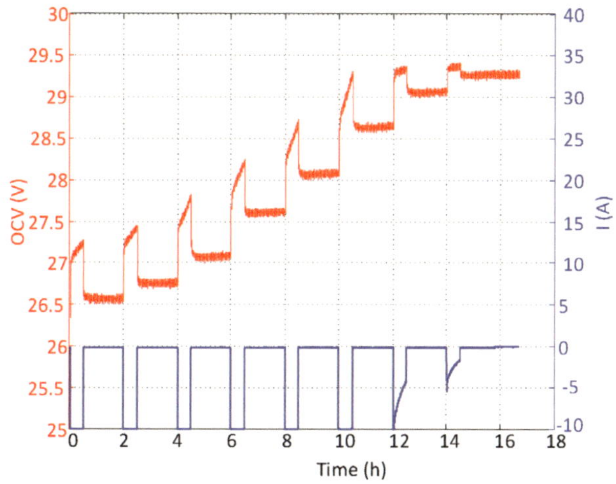

Figure 3.
Charge test result.

Discharge results		Charge results	
OCV	SOC (%)	OCV	SOC (%)
29.40	100	29.28	90.27
28.77	90	29.06	87.64
28.23	80	28.65	80
27.79	70	28.09	70
27.40	60	27.62	60
26.89	50	27.08	50
26.65	40	26.76	40
26.49	30	26.57	30
26.33	20	26.33	20

Table 2.
OCV values at each test point.

2.2 Modeling of the battery-pack's complex impedance

To carry out EIS tests, an impedance analyzer is generally used. This device generates a frequency sweep signal and measures the voltage and current in each test point. As a result, the complex impedance is calculated. Because most of commercial impedance analyzer generates AC signals less than 100 mA (suitable for cell testing), in this work, this signal is amplified and controlled by means of the experimental test bench deeply explained in [26].

EIS tests have been performed at different SOC values (20, 40, 60, 80, and 90% SOC) to analyze the effects of these SOC variations. The frequency sweep has been set from 1 mHz to 5 kHz (typically test range), with an AC ripple of 5 A (10% I_{max}). **Figure 5** shows the Nyquist plots of the EIS results, which has been used to analyze the impedance behavior of the tested battery module. In this graph, the real part of the complex impedance (Z') is represented along the x-axis and the imaginary part (Z") along the y-axis. The capacitive behavior corresponds to negative values of Z" and the inductive behavior to the positive ones. In this way, it is easy to identify the parameters of the electrical circuit. According to EIS results, the impedance

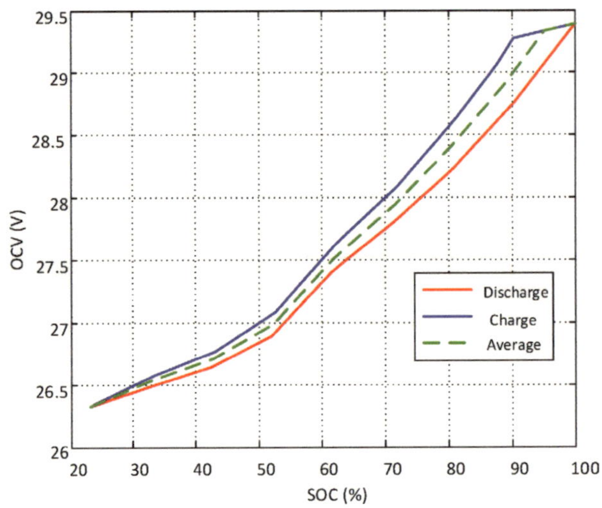

Figure 4.
OCV-SOC characteristic.

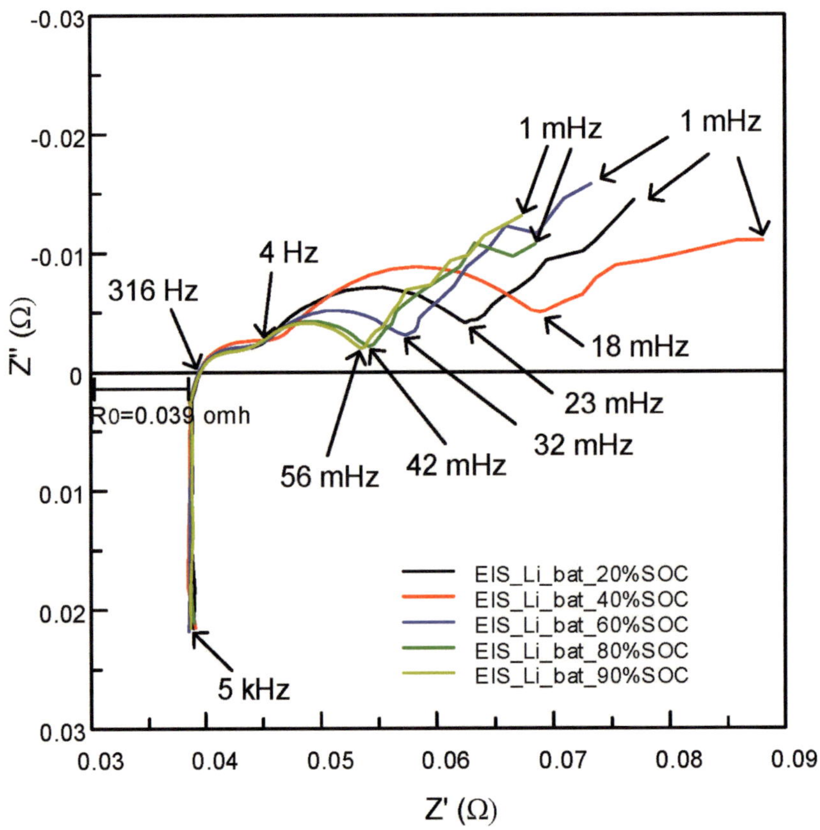

Figure 5.
EIS tests result.

of the pack tested shows a capacitive behavior from 1 mHz to 316 Hz. From this value, the pack impedance corresponds to an ideal inductance. The differences in these plots reflect that SOC variations affect the capacitive behavior from 1 mHz to 4 Hz. For low frequencies, the Nyquist plots show that both real (Z') and imaginary

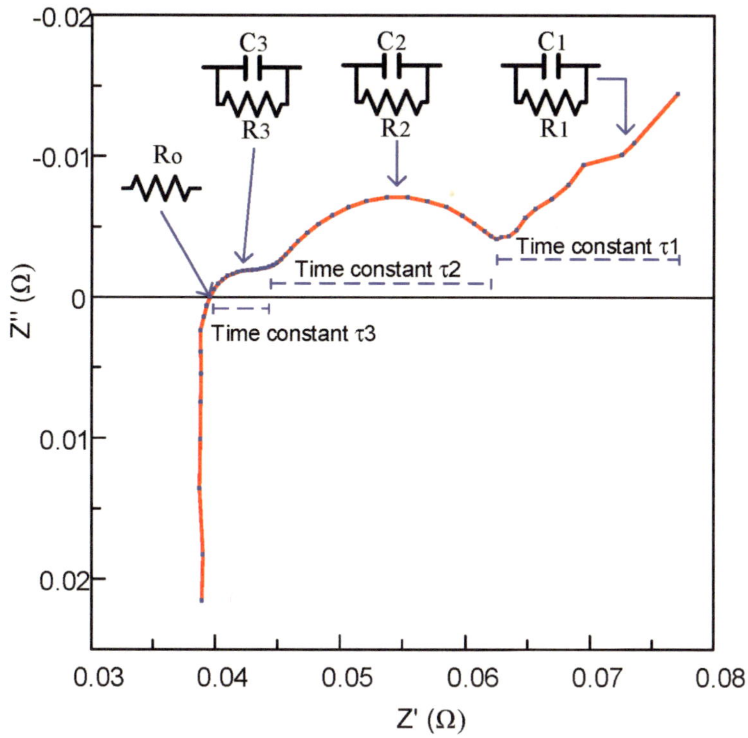

Figure 6.
Time constants associated with pack impedance.

(Z”) parts of the impedance decrease with frequency, drawing a line with a slope of almost 45°. From 18 mHz to 4 Hz, the plots present semicircular shapes, whose diameter diminishes with increasing SOC. For medium frequencies (4 to 316 Hz), the impedance corresponds to a semicircle of constant diameter. To simulate these capacitive behaviors, several *RC* networks connected in series can be used [20, 21]. The value of equivalent resistance of the battery-pack is $R_o \cong 39$ mΩ.

To analyze the influence of the impedance components on the dynamic response of the battery-pack, **Figure 6** shows the EIS result at 40% SOC. The results analysis allows to associate the most relevant time constants with the impedance behavior of the battery-pack. Most of the dynamic applications of the batteries (load/frequency control or renewable generation support) have their time constants from 1 mHz to 316 Hz; for this reason, the inductive behavior can be neglected. In this frequency range, the *RC* networks that reproduce the impedance behavior can be used to determine the different time constants that affect the dynamic response of the tested battery-pack. These time constants $(\tau 1, \tau 2, \tau 3)$ are calculated from EIS test results and presented in **Table 3**.

2.3 Hybrid model of the battery-pack

As a result of these combined time- and frequency-domain tests, an electrical circuit has been determined. Also, the model includes an integration current SOC estimator, to guarantee that the parameters of the electrical circuit simulate the dynamic behavior of the battery-pack for different SOC conditions, as shown in **Figure 7**. The inputs of the model are the initial value of SOC (SOC_0), the battery-pack capacity (Cn), and the current (i_{pack}). The output corresponds to the voltage response of the battery-pack (u_{pack}), which is calculated by Eq. (2), where uRo, $uc1$,

Element	20% SOC	40% SOC	60% SOC	80% SOC	90% SOC
Ro (Ω)	0.039	0.039	0.039	0.039	0.039
τ1(C1//R1) (s)	23.40	18.43	15.71	11.39	10.92
τ2(C2//R2) (s)	0.295	0.188	0.136	0.116	0.106
τ3(C3//R3) (s)	0.0028	0.0030	0.0030	0.0029	0.0028

Table 3.
Battery-pack impedance parameters.

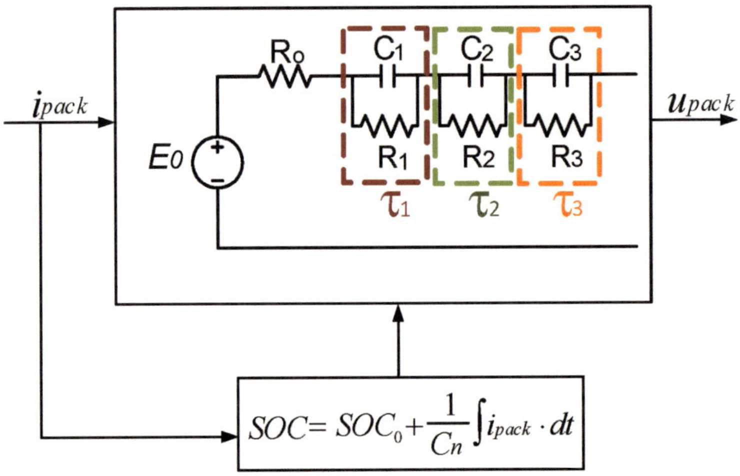

Figure 7.
Battery-pack model.

uc2, and *uc3* simulate the voltage response of the ohmic resistance and each *RC* network, respectively:

$$u_{pack}(SOC,t) = Eo(SOC,t) - u_{Ro}(SOC,t) - u_{C1}(SOC,t) - u_{C2}(SOC,t) - u_{C3}(SOC,t) \quad (2)$$

where:

$$u_{Ro}(SOC,t) = R_o(SOC) \cdot ipack(t) \quad (3)$$

$$u_{C1}(SOC,t) = \int \frac{1}{C_1(SOC)} \cdot \left(ipack(t) - \frac{u_{C1}(SOC)}{R_1(SOC)} \right) \cdot dt \quad (4)$$

$$u_{C2}(SOC,t) = \int \frac{1}{C_2(SOC)} \cdot \left(ipack(t) - \frac{u_{C2}(SOC)}{R_2(SOC)} \right) \cdot dt \quad (5)$$

$$u_{C3}(SOC,t) = \int \frac{1}{C_3(SOC)} \cdot \left(ipack(t) - \frac{u_{C3}(SOC)}{R_3(SOC)} \right) \cdot dt \quad (6)$$

3. Hybrid model validation

To verify the accuracy and reliability of the proposed model to simulate the battery-pack behavior, hardware-in-the-loop simulations are used. HIL is a widely used experimental technique to reproduce the real conditions of physical applications [27, 28], using lab devices such as electronic loads, power sources, sensors, and data acquisition systems (in the case of electric applications). To perform the HIL simulation of the battery-pack, the experimental setup shown in [26] is used. In this test bench, the load regime is simulated by means of a MATLAB/Simulink model (software simulation). This current signal is used to control (by DSpace control system) the output signal of an electronic load and a power source connected in parallel to reproduce the charging/discharging cycles. This configuration is called hardware simulation; in this way, real devices are used to test the battery-pack. A control schema of the HIL simulation is shown in **Figures 8** and **9** that presents a picture of the test bench.

Three simulations of the battery-pack performance under dynamic regimes associated with distribution grids operation have been analyzed. The first one corresponds to a load frequency control application (LFC), which is related to grid frequency control, with typical time constants ranging from 0.2 ms to 10 s. The second one reproduces the dynamic voltage support (DVS) of a renewable energy source during 110 s. The simulated models of these load regimes are based on the operation of a hybrid ac/dc microgrid presented in [29]. Finally, the third one simulates the performance of an energy support device uninterruptible power supply (UPS). The time duration of this energy support is less than 30 min; and to

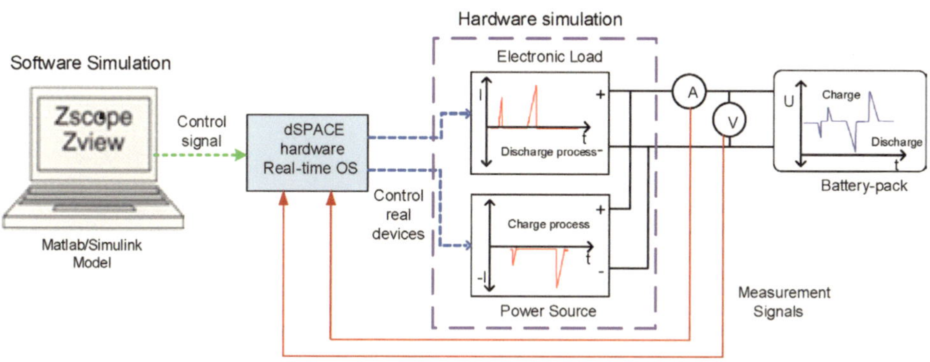

Figure 8.
HIL simulation control setup.

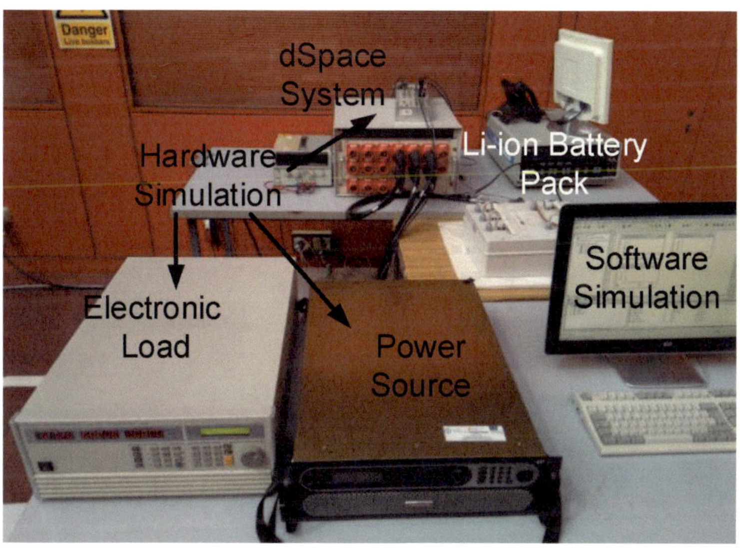

Figure 9.
Test bench picture.

test the transient response, different current steps have been simulated. The current profiles associated with these applications are shown in **Figures 10–12**. These current signals are used as the input signals of the HIL simulations. The simulations have been performed at different SOC values to check the model response under SOC variations. **Figures 13–18** present the comparison of the voltage response of the hybrid model (U_{model}) and the voltage measurement at battery-pack terminals (U_{bpack}) during the HIL simulations.

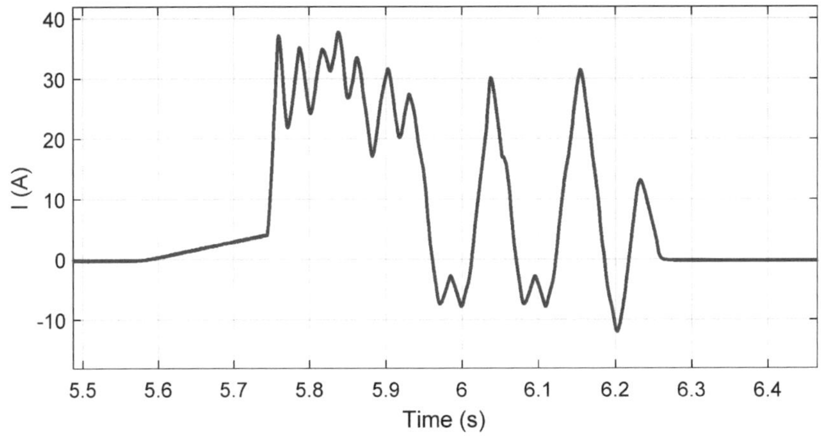

Figure 10.
Current profile of LFC simulation.

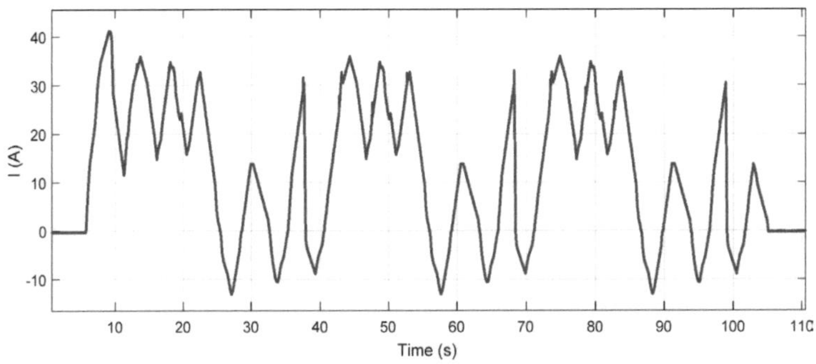

Figure 11.
Current profile of DVS simulation.

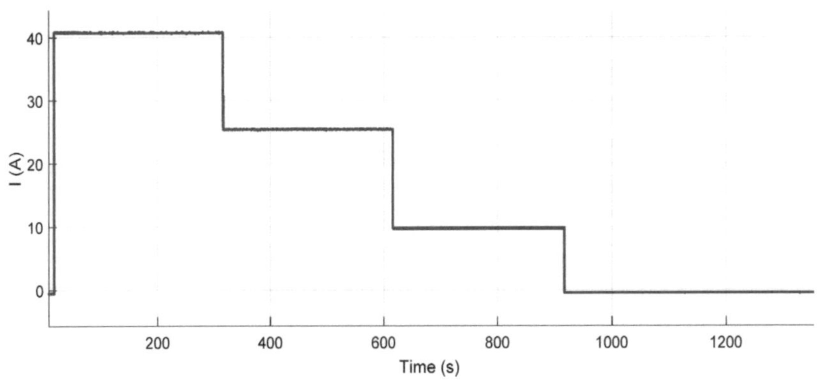

Figure 12.
Current profile of UPS simulation.

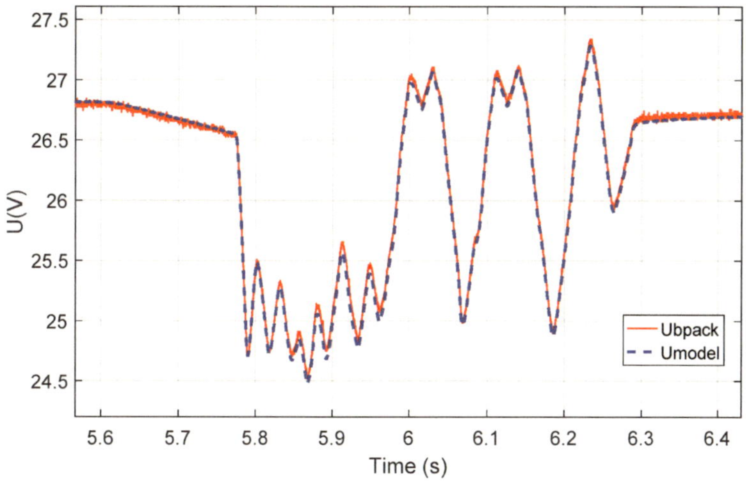

Figure 13.
Simulation results at LFC (48% SOC).

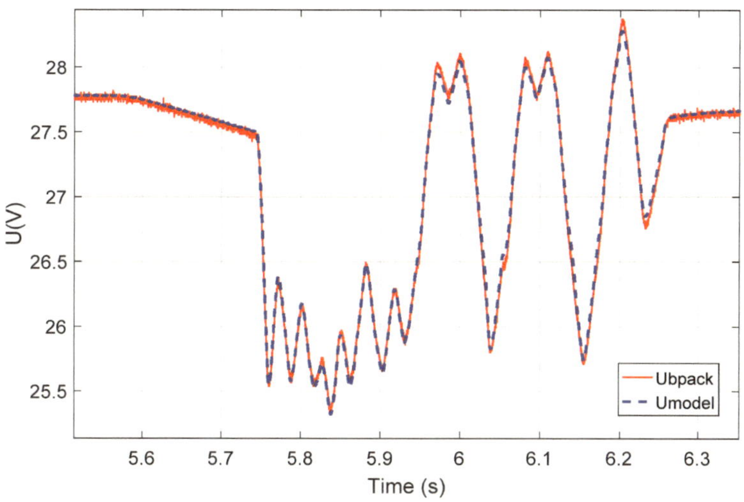

Figure 14.
Simulation results at LFC (72% SOC).

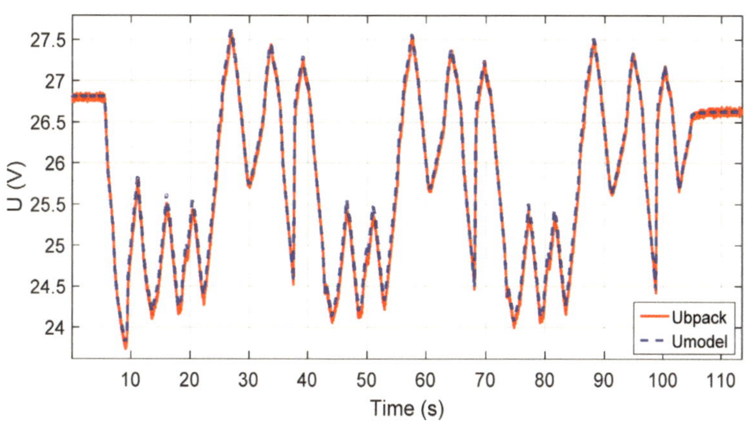

Figure 15.
Simulation results at DVS (45% SOC).

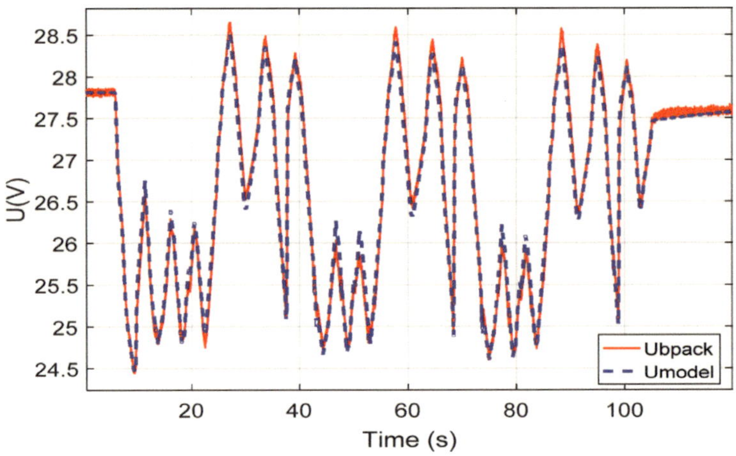

Figure 16.
Simulation results at DVS (73% SOC).

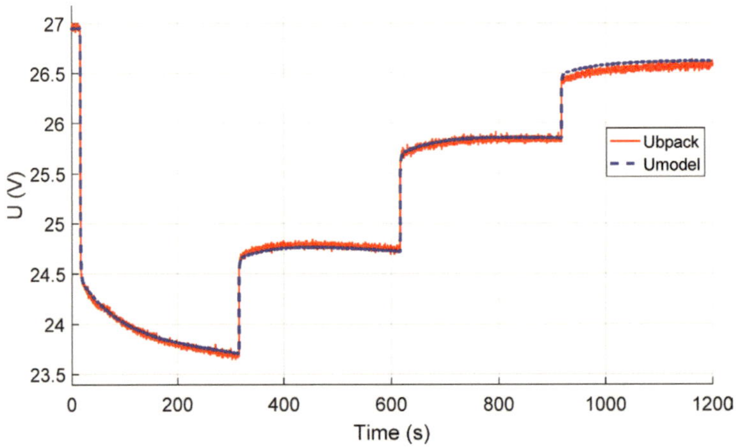

Figure 17.
Simulation results at UPS (53% SOC).

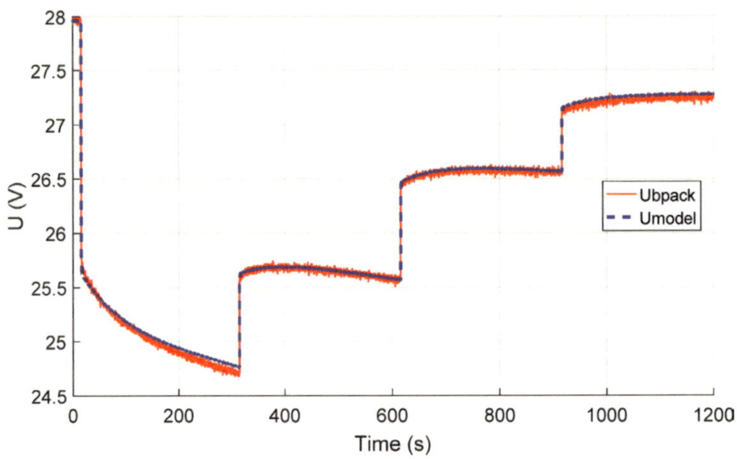

Figure 18.
Simulation results at UPS (78% SOC).

The validations tests show that the hybrid model simulates the battery-pack behavior with high accuracy in all cases analyzed. The maximum errors for each simulation are 0.28% (LFC), 0.40% (DVS), and 0.23% (UPS).

4. Conclusions

This chapter presents a hybrid modeling procedure of Li-ion battery-packs which is able to simulate the dynamic behavior associated with electric grid applications. The parameters of an electrical circuit have been calculated from experimental results of current interruption and EIS tests. The active behavior of the battery-pack has been simulated by a voltage source, and the impedance reflects the electrochemical processes by means of three *RC* networks (which correspond to three different time constants) and an ohmic resistance. The experimental procedure has been performed at the whole battery-pack in order to include the interactions of battery cells and BMS components.

To reproduce the battery-pack behavior under high dynamic applications of distribution grids, a hardware-in-the-loop platform has been used. Three different cases (load frequency control, dynamic voltage support, and uninterruptible power supply) at different SOC conditions have been simulated. The validation results show that the hybrid model reproduces the dynamic behavior of the battery-pack with high accuracy in all cases analyzed.

Conflict of interest

The authors declare no conflict of interest.

Author details

Sandra Castano-Solis[1]*, Daniel Serrano-Jiménez[2], Jesús Fraile-Ardanuy[3], David Jiménez-Bermejo[3] and Javier Sanz-Feito[4]

1 ETS Ingeniería y Diseño Industrial, Universidad Politécnica de Madrid, Madrid, Spain

2 Electronic Engineering and Computation Science Department, Universidad de Mondragón, Guipúzcua, Spain

3 ETSI Telecomunicación, Universidad Politécnica de Madrid, Madrid, Spain

4 Electrical Engineering Department, Universidad Carlos III de Madrid, Madrid, Spain

*Address all correspondence to: sp.castano@upm.es

References

[1] Mohamad F, Teh J, Lai C-M, Chen L-R. Development of energy storage systems for power network reliability: A review. Energies. 2018;**11**(9):2278

[2] Hill CA, Such MC, Chen D, Gonzalez J, Grady WM. Battery energy storage for enabling integration of distributed solar power generation. IEEE Transactions on Smart Grid. 2012;**3**(2):850-857

[3] Luo X, Wang J, Dooner M, Clarke J. Overview of current development in electrical energy storage technologies and the application potential in power system operation. Applied Energy. 2015;**137**:511-536

[4] Abada S, Marlair G, Lecocq A, Petit M, Sauvant-Moynot V, Huet F. Safety focused modeling of lithium-ion batteries: A review. Journal of Power Sources. 2016;**306**:178-192

[5] Li S, Mi CC, Zhang M. A high-efficiency active battery-balancing circuit using multiwinding transformer. IEEE Transactions on Industry Applications. 2013;**49**:198-207

[6] Li J, Greye B, Buchholz M, Danzer MA. Interval method for an efficient state of charge and capacity estimation of multicell batteries. Journal of Energy Storage. 2017;**13**:1-9

[7] Hu Y, Yurkovich S, Guezennec Y, Yurkovich BJ. A technique for dynamic battery model identification in automotive applications using linear parameter varying structures. Control Engineering Practice. 2009;**17**(10):1190-1201

[8] Sikha G, White R, Popov B. A mathematical model for a lithium-ion battery/electrochemical capacitor hybrid system. Journal of the Electrochemical Society. 2005;**152**(8):A1682-A1693

[9] Smith K, Rahn C, Wang C. Model-based electrochemical estimation and constraint management for pulse operation of lithium ion batteries. IEEE Transactions on Control Systems Technology. 2010;**18**(3):654-663

[10] Shafiei A, Momeni A, Williamson S. Battery modeling approaches and management techniques for plug-in hybrid electric vehicles. In: Proceedings of Vehicle Power and Propulsion Conference, VPPC'11. 2011. pp. 1-5

[11] Abu-Sharkh S, Doerffel D. Rapid test and non-linear model characterization of solid-state lithium-ion batteries. Journal of Power Sources. 2004;**130**:266-274

[12] He H, Xiong R, Zhang X, Sun F, Fan J. State-of-charge estimation of the lithium-ion battery using an adaptive extended Kalman filter based on an improved Thevenin model. IEEE Transactions on Vehicular Technology. 2011;**60**(4):1461-1469

[13] Purvins A, Papaioannou IT, Debarberis L. Application of battery-based storage systems in household-demand smoothening in electricity-distribution grids. Energy Conversion and Management. 2013;**65**:272-284

[14] Chan HL, Sutanto D. A new battery model for use with battery energy storage systems and electric vehicles power systems. In: Proceedings of Power Eng Soc Winter Meet. 2000. pp. 470-475

[15] He H, Xiong R, Guo H, Li S. Comparison study on the battery models used for the energy management of batteries in electric vehicles. Energy Conversion and Management. 2012;**64**:113-121

[16] Zheng Y, Gao W, Ouyang M, Lu L, Zhou L, Han X. State-of-charge

inconsistency estimation of lithium-ion battery pack using mean-difference model and extended Kalman filter. Journal of Power Sources. 2018;**383**:50-58

[17] El Din MS, Abdel-Hafez MF, Hussein AA. Enhancement in Li-ion battery cell state-of-charge estimation under uncertain model statistics. IEEE Transactions on Vehicular Technology. 2016;**65**(6):4608-4618

[18] Andre D, Meiler M, Steiner K, Walz H, Soczka-Guth T, Sauer DU. Characterization of high-power lithium-ion batteries by electrochemical impedance spectroscopy. II: Modelling. Journal of Power Sources. 2011;**196**:5349-5356

[19] Castano S, Gauchia L, Sanz J. Effect of packaging on Supercapacitors string modeling: proposal of a functional unit defined around the balancing circuit. IEEE Transactions on Components, Packaging and Manufacturing Technology. 2013;**3**(8):1390-1398

[20] Buller S, Thele M, De Doncker RWAA, Karden E. Impedance-based simulation models of supercapacitors and Li-ion batteries for power electronic applications. IEEE Transactions on Industry Applications. 2005;**41**(3):742-747

[21] Karden E, Buller S, De Doncker RW. A frequency-domain approach to dynamical modeling of electrochemical power sources. Electrochimica Acta. 2002;**47**(13):2347-2356

[22] Castano-Solis S, Serrano-Jimenez D, Fraile-Ardanuy J, Sanz-Feito J. Hybrid characterization procedure of Li-ion battery packs for wide frequency range dynamics applications. Electric Power Systems Research. 2019;**166**:9-17

[23] Castano-Solis S, Serano-Jimenez D, Gauchia L, Sanz J. The influence of BMSs on the characterization and modeling of series and parallel Li-ion packs. Energies. 2017;**10**(3):273

[24] Jossen A. Fundamentals of battery dynamics. Journal of Power Sources. 2006;**154**(2):530-538

[25] Dai H, Wei X, Sun Z, Wang J, Gu W. Online cell SOC estimation of Li-ion battery packs using a dual time-scale Kalman filtering for EV applications. Applied Energy. 2012;**95**:227-237

[26] Castano-Solis S, Gauchia L, Serrano Jimenez D, Sanz Feito J. Off-the-shelf and flexible hybrid frequency and time domain experimental architecture setup for electrochemical energy modules testing under realistic operating conditions. IEEE Transactions on Energy Conversion. 2017;**32**(2):620-627

[27] Huerta F, Tello RL, Prodanovic M. Real-time power-hardware-in-the-loop implementation of variable-speed wind turbines. IEEE Transactions on Industrial Electronics. 2017;**64**(3):1893-1904

[28] Gauchia L, Sanz J. A per-unit hardware-in-the-loop simulation of a fuel cell/battery hybrid energy system. IEEE Transactions on Industrial Electronics. 2010;**57**(4):1186-1194

[29] Liu X, Wang P, Loh PC, Hybrid A. AC/DC microgrid and its coordination control. IEEE Transactions on Smart Grid. 2011;**2**(2):278-286